RADIOCHEMISTRY—
THEORY AND EXPERIME

The Wykeham Science Series

The Author

T. A. H. PEACOCKE taught chemistry at St. John's School, Leatherhead and at Charterhouse over a period of more than 35 years up to 1975. He has done much original work during the last 25 years on the use of radioactive isotopes in chemistry teaching, has published original work on the Szilard–Chalmers reaction and has carried out unpublished work on methods for separating the lanthanides, using tracers. He is the author of *Atomic and Nuclear Chemistry* (1967) and *Small Scale Experimental Chemistry* (1960, new edition 1972). He has made films and broadcasts on chemistry and radiochemistry for schools.

RADIOCHEMISTRY—THEORY AND EXPERIMENT

T. A. H. PEACOCKE

Formerly Head of the Science Department of St John's School, Leatherhead

WYKEHAM PUBLICATIONS (LONDON) LTD
(A member of the Taylor & Francis Group)
LONDON 1978

First published 1978 by Wykeham Publications (London) Ltd.

ISBN 0 85109 690 5 (Paper)

Printed in Great Britain by Taylor & Francis (Printers) Ltd.
Rankine Road, Basingstoke, Hants. RG24 0PR

Distribution:
UNITED STATES OF AMERICA AND CANADA
Crane, Russak & Company, Inc.,
347 Madison Avenue, New York, N.Y. 10017, U.S.A.

AUSTRALIA, NEW ZEALAND AND FAR EAST
Australian and New Zealand Book Co. Pty. Ltd.,
P.O. Box 459, Brookvale, N.S.W. 2100.

JAPAN
Kinokuniya Book-Store Co. Ltd.,
17–7 Shinjuku 3 Chome, Shinjuku-ku, Tokyo 160–91, Japan.

INDIA, BANGLADESH, SRI LANKA AND BURMA
Arnold-Heinemann Publishers (India) Pvt. Ltd.,
AB-9, First Floor, Safdarjang Enclave, New Delhi-110016.

ALL OTHER TERRITORIES
Taylor & Francis Ltd., 10–14 Macklin Street, London WC2B 5NF.

Contents

Some radiochemical experiments

1. The statistics of radiochemical measurements.
2. Growth and decay.
3. To separate uranium 'X' from uranium.
4. To separate ^{208}Tl from thorium nitrate and to determine the half-life.
5. To separate ^{137m}Ba from ^{137}Cs and to determine the half-life.
6. To determine the half-life of a long lived radionuclide (^{238}U), using a calibrated G–M tube.
7. To determine the half-life of a short lived radionuclide by the integral, or 'ratio' method.
8. To resolve a decay curve with two components.
9. To plot a γ-spectrum.
10. To plot a β-absorption curve and to determine the maximum β-energy.
11. To measure the resolving time of a counter assembly by:
    (*a*) the method of paired sources;
    (*b*) successive additions of a given nuclide of long half-life;
    (*c*) extrapolation of a known decay curve.
12. An introduction to liquid scintillation counting.
13. To determine the thermal neutron flux in a paraffin block containing a neutron source.
14. Neutron activation analysis:
    (*a*) to detect impurities in a compound by reactor irradiation;
    (*b*) to determine the percentage of manganese in a specimen of 'Dural' by radioactivation analysis, using a neutron source.
15. To determine the percentage retention of ^{128}I in the organic phase after neutron irradiation of iodoethane.
16. To determine the distribution coefficient for iodine between water and 1,1,1-trichloroethane.
17. To investigate the solubility of lead chloride in hydrochloric acid solutions using ^{212}Pb as a tracer, and to determine the value of the stability constant for the $PbCl_4^{2-}$ complex ion.
18. To plot a solubility curve for lead iodide using ^{212}Pb as a tracer.
19. To determine the stability constant for the silver ammine complex ion.
20. To investigate the equilibrium between iodide ions and molecular iodine in aqueous solution.
21. To determine the solubility product of silver ethanoate (acetate) at a known temperature and to investigate the effect of 'ionic strength' on solubility.
22. To determine the transport number of the Ag^+ in silver nitrate solution.
23. To determine phosphate labelled with ^{32}P by radiochemical titration with barium nitrate.

Appendices

To Constance

Preface

The writing of this book is the result of some twenty-five years' work in this field. It is intended for use in first and second year courses in universities and polytechnics, and in the final year of schools. It contains both theory and experiment and the chronological arrangement of the early chapters may also be of use to those who wish to study the original work in radioactivity, but may not have the time or the facilities to consult the original papers. Those who wish to do so will find all the necessary references.

The author feels strongly that the student of this subject should know how it developed. For example, the discovery of radioactivity itself and of nuclear fission were both quite unexpected and were planned to give quite different results. The research worker of the future who studies the work of the pioneers, who often worked with comparatively primitive apparatus, will come to realize that, if he wishes to advance the frontiers of knowledge, he too must be ready for the unforeseen observation, and must have an infinite capacity for taking pains.

The first three chapters present the subject in historical order, so that each new discovery fits into the general pattern and so makes a fascinating story. As far as space allows, brief notes have been included on the character and personality of the great scientists involved, so that they may be seen as real people and not as mere names. Much of this could be followed by an interested layman. Chapter 4 gives some account of the discovery and properties of the actinides or second rare earth series. Chapters 5–9 deal with the necessary theory, the solutions of the more complex equations being given in the appendix. Chapters 10 and 11 give some account of practical applications with simple illustrative experiments and Chapter 12 deals with radiological safety.

Certain chemical experiments, such as exchange reactions, can be carried out only with the aid of isotopes and many others can be enriched and simplified by their use. Students find work with radioisotopes both exciting and stimulating, and respond to the challenges imposed by the techniques required. Thirty-five experiments have been included with full working details. All the necessary precautions and safe handling techniques are fully described, but the quantities of radioactive material recommended for use are well within the safety tolerances, as given, for example, in A.M.65 published by the DES. All the experiments have been fully tested in the laboratories of two schools by advanced sixth-form boys, often carrying them out as projects, and in most cases they can be done with G-M tubes. Many of the experiments have been developed by the author and to the best of his knowledge are original. There are still a number of other aspects of chemistry where radioisotopes could be used with advantage, and the author hopes that this book may help others to find the great interest which he himself has enjoyed.

Units and nomenclature. S.I. units have, in general, been used throughout the book, with the exception of electron volts and rads. In the case of electron volts, the joule equivalent has been included in parentheses. Barns are quoted in m^2 or cm^2 according to convenience, i.e. when using neutron fluxes, which are normally quoted in cm^{-2} barns are quoted in cm^2, following usual radiochemical practice.

Modern chemical nomenclature has been used, but the conventional names of common substances have also been given for those who prefer them.

T. A. H. PEACOCKE

Acknowledgments

I am indebted to many people who have helped materially with the preparation of this book: Professor G. N. Walton, Dr J. M. Blatchly, Dr D. F. Shaw, and Mr M. Berry who have read the manuscript and made many helpful suggestions: Professor F. W. Shotton, F.R.S. for information on the current methods of radiocarbon dating: Dr H. D. Evans for supplying the figures for Table 12.2.: Dr E. Glueckauf, F.R.S. for information on the original work in the Szilard–Chalmers reaction: Professor A. L. Odell for permission to use Experiments 23 and 30: Dr J. A. Bassham for providing the photographs on pages 236 and 237: Dr J. R. Catch for advice with Experiment 28 and Mr D. J. Angwin for advice with Experiment 31: Dr R. A. Wright for assistance with Experiment 25: Mr G. Ullyott for help in the presentation of some parts of the appendix: my many VI form students both at Charterhouse and at St John's for their help in testing all the experiments: the Reactor Manager and Staff of the University of London Reactor for facilities: Miss J. C. Foxwell for typing a large part of the manuscript and last, but by no means least, the Publishers, Wykeham Publications (London) Ltd for their help and forbearance in the preparation of the manuscript.

I wish to thank the U.K.A.E.A. for providing Figures 3.1–3.5, the Oxford Colleges Admissions Office, the Oxford and Cambridge Schools Examination Board and the Nuffield Foundation for permission to use certain questions, Messrs Prentice-Hall, The Oxford University Press, The Cambridge University Press, The McGraw-Hill Book Co., Messrs. John Wiley and Sons and Dr H. R. Hulme for kindly giving permission to reproduce certain figures.

Lastly, the author wishes to make it clear that he alone accepts full responsibility for any errors or omissions.

The Periodic Classification (with atomic numbers)

Group / *Period*	IA	IIA	IIIA															IVA	VA	VIA	VIIA	VIII			IB	IIB	IIIB	IVB	VB	VIB	VIIB	
1	1 H																															2 He
2	3 Li	4 Be																									5 B	6 C	7 N	8 O	9 F	10 Ne
3	11 Na	12 Mg																									13 Al	14 Si	15 P	16 S	17 Cl	18 Ar
4	19 K	20 Ca	21 Sc															22 Ti	23 V	24 Cr	25 Mn	26 Fe	27 Co	28 Ni	29 Cu	30 Zn	31 Ga	32 Ge	33 As	34 Se	35 Br	36 Kr
5	37 Rb	38 Sr	39 Y															40 Zr	41 Nb	42 Mo	43 Tc	44 Ru	45 Rh	46 Pd	47 Ag	48 Cd	49 In	50 Sn	51 Sb	52 Te	53 I	54 Xe
6	55 Cs	56 Ba	57 La	58 Ce	59 Pr	60 Nd	61 Pm	62 Sm	63 Eu	64 Gd	65 Tb	66 Dy	67 Ho	68 Er	69 Tm	70 Yb	71 Lu	72 Hf	73 Ta	74 W	75 Re	76 Os	77 Ir	78 Pt	79 Au	80 Hg	81 Tl	82 Pb	83 Bi	84 Po	85 At	86 Rn
7	87 Fr	88 Ra	89 Ac	90 Th	91 Pa	92 U	93 Np	94 Pu	95 Am	96 Cm	97 Bk	98 Cf	99 Es	100 Fm	101 Md	102 No	103 Lr	104 –	105 –													

s-block: 3 Li, 4 Be (Groups IA, IIA)

d-block: 20 Ca, 21 Sc; d-block: 22 Ti – 30 Zn

p-block: 5 B – 10 Ne

Lanthanides — f-block: 57 La – 71 Lu

Actinides: 89 Ac – 103 Lr

1. The early investigations

1.1. *The discovery of radioactivity*

Like so many great scientific discoveries, this one resulted from a chance observation and a fortunate chain of circumstances. Early in 1896 Henri Becquerel was investigating the fluorescence of uranium salts. His father had already done some work in this field. Röntgen had discovered the remarkable properties of X-rays the previous year, and there seemed to be a connection between X-rays and fluorescence. Becquerel wished to find out whether the fluorescence of uranium salts was accompanied by an emission of a penetrating radiation like X-rays. On 20 February he wrapped a photographic plate in two thicknesses of black paper, covered this with a layer of crystalline potassium uranyl sulphate [$K_2UO_2(SO_4)_2 . 2H_2O$], a substance he had himself prepared in 1880, and exposed the whole to the sun for several hours. On developing the plate he found a blackening in the area beneath the uranium salt. He repeated his experiment, interposing a coin between the salt and the plate, and obtained an image of the coin (see Experiment 35).

He intended to confirm his results on 26 February and again wrapped his photographic plate in black paper and sprinkled it with crystals of the same uranium salt. The weather, however, turned misty and the sunshine was so feeble that Becquerel put the plate, still covered with the uranium salt, in a drawer. By 1 March the weather was still no better, so he developed the plate, expecting to find a very feeble image. To his intense surprise the blackening produced was much more marked than before.

Becquerel repeated his experiments in the dark, and he even prepared a specimen of potassium uranyl sulphate in the dark. In every experiment he obtained exactly the same result. He tried a variety of uranium compounds and even the metal itself which had recently been prepared by Moissan. Many of these compounds did not show fluorescence, but they all fogged the plate, uranium metal producing the most marked result of all.

He next interposed various materials such as aluminium sheet and thin sheets of glass and paper between his uranium compound and the plate, and found that these objects were penetrated to varying degrees. He showed that the rays emitted could discharge an electroscope and concluded that they were similar to Röntgen rays [1]. Although Becquerel continued to experiment in this field for some years, the next major advance was made by Marie Curie, a student in his laboratory. She first proved that the intensity of the radiation per unit mass was proportional to the percentage of uranium present in the compound. Next she examined a number of different elements and discovered the radioactivity of thorium; it was Marie Curie who named the property radioactivity.

1.2. *The isolation of radium*

Her next discovery was that the radioactivity of the uranium ore pitchblende was about eight times as strong as would have been expected from its uranium content. She concluded that it contained small quantities of some new element. She and her husband Pierre separated all the elements in pitchblende and measured the radioactivity of each. They found that the greatest intensity was concentrated in two fractions, one containing barium and the other containing bismuth. They concluded that each fraction contained a new element, and they named these elements radium and polonium.

To isolate these two elements it would be necessary to start with a large quantity of pitchblende, which was a costly ore, so, reasoning that the polonium and radium would be left in the residues after the extraction of the uranium, they obtained a gift from the Austrian Government of one ton of this residue. Pitchblende contains approximately $0{\cdot}2$ g of radium and 4×10^{-5} g of polonium per tonne. To separate the radium it was precipitated along with barium as sulphate. The barium sulphate containing the radium was converted to the chloride and the radium was separated by fractional crystallization, radium chloride being less soluble than barium chloride in the ratio of $1 : 4{\cdot}5$. It took the Curies four years to separate pure radium chloride, and they eventually obtained $0{\cdot}1$ g. Besides its intense radioactivity, it glowed in the dark with a bluish light. It gave out heat, approximately $0{\cdot}1\ \mathrm{W\,g^{-1}}$.

Next, radium metal was isolated by the electrolysis of an aqueous solution of radium chloride with a mercury cathode, the mercury being distilled off afterwards. Radium metal is very reactive and turns black in air, probably owing to the formation of a nitride. The relative atomic mass of radium was determined and found to be about 225. Subsequently Madame Curie obtained the value 226·5. The latest figure is 226·05 [2].

The polonium was precipitated with bismuth but, owing to the very small proportion of polonium in pitchblende, the Curies failed to isolate it. It was, however, obtained by W. Marckwald in 1902 as a thin deposit on a bismuth plate placed in a solution of bismuth chloride obtained from pitchblende.

A third radioactive element was discovered in the pitchblende residues by A. Debierne, who was working with the Curies. It was precipitated with the lanthanides (or first rare-earth series) and was called actinium, although at that time it was not possible to isolate it. This is not surprising since the proportion of actinium in pitchblende is about 1 part in 10^{10} [3].

1.3. *The nature of the radiations*

In 1899, Ernest Rutherford began a study of the rays emitted by radioactive bodies. He found these rays to be of a complex nature. First, alpha rays: these were particles which were stopped by very thin sheets of aluminium foil and deflected very slightly by magnetic fields, the direction of the deflection indicating a positive charge. Secondly, beta rays: these were also particles but of much smaller mass than the alpha rays, penetrating considerable thicknesses of aluminium foil and up to 30 cm of air. They were deflected fairly easily by magnetic fields, the direction of the deflection indicating a negative charge. Lastly, gamma rays: these penetrated several cm of aluminium and several metres of air, the intensity in air falling off roughly as the square of the distance. They were completely undeflected by magnetic fields.

1.4. *Growth and decay*

The experiments of Marie Curie had shown that the radioactivity of uranium was a property of uranium *atoms*, the radioactivity of a given sample of a uranium compound depending only on the number of uranium atoms present in it. It therefore seemed most unlikely that the activity of uranium could in any way be diminished by chemical or physical means. In 1900, however, Sir William Crookes observed that, if ammonium carbonate solution was added to a solution of a uranium salt until the precipitate first formed was redissolved, it left a tiny almost invisible residue. This residue contained practically the whole of the activity of the original uranium, which meanwhile had become almost inactive. The activity was measured by the photographic method. Thus by a single chemical operation Crookes had been able to remove the activity from the uranium [4]. Results of a similar nature were obtained later by Becquerel. This active residue was called by Crookes uranium 'X' (Experiment 3). Uranium 'X' and the original uranium were set aside for a year. When they were re-examined, the uranium 'X' appeared to have lost the whole of its activity, whilst the uranium had completely regained its activity. The loss of activity was thus shown to be purely temporary. It was fortunate that the activity was measured by the photographic method, i.e. spreading the compound on black paper covering a photographic plate, for, had it been measured electrically with the aid of an electroscope, the alpha activity which is retained by the uranium would have masked the loss of beta activity and the effect would not have been observed.

In 1902 Rutherford and Frederick Soddy made a similar observation with thorium compounds. They found that if ammonia was added to a thorium solution, the thorium was precipitated, but a large amount of the activity was left behind in the filtrate, which was chemically free from thorium. After the ammonium salts had been driven off by ignition, the residue was found to contain more than half of the activity of the original thorium, and in proportion to its weight to be several thousand times more active. This active constituent was named thorium 'X'. When the activity was examined about a month later, it was found that thorium 'X' was inactive, while the thorium had regained its original activity. Rutherford then measured the rates of gain and loss of activity of the thorium and the thorium 'X' respectively, and so obtained growth and decay curves (Figure 1.1 and Experiment 2). He found that the decay followed the exponential law:

$$A = A_0 \exp(-\lambda t) \tag{1.1}$$

where A_0 was the original activity, A the activity after time t, and λ a constant now known as the disintegration constant.

Equation 1.1 can be written

$$\ln A = -\lambda t + \ln A_0 \tag{1.2}$$

Differentiating with respect to t:

$$-\frac{\mathrm{d}A}{A} = \lambda \mathrm{d}t$$

or:

$$-\frac{\mathrm{d}A}{\mathrm{d}t} = \lambda A \tag{1.3}$$

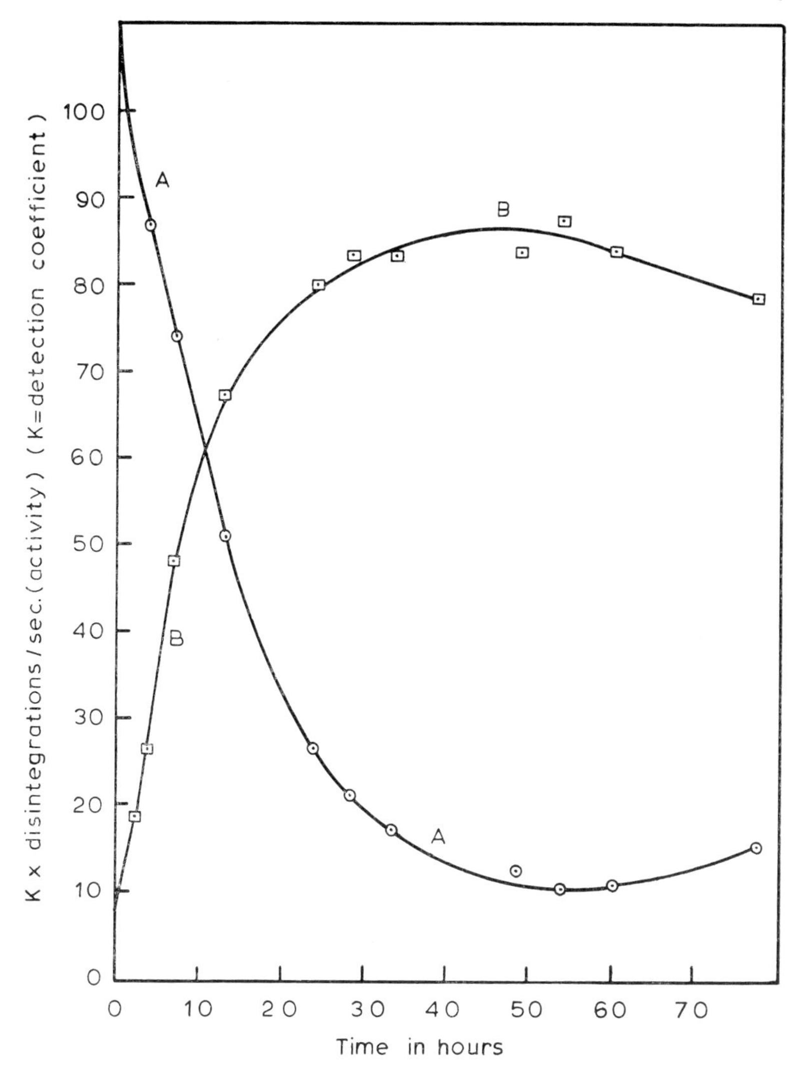

Figure 1.1. Growth and decay. When ammonia solution (carbonate free) is added to thorium solution, $Th(OH)_4$ is precipitated and carries down ^{228}Ac, ^{212}Pb, ^{212}Bi and ^{208}Tl as hydroxides, but leaves the radium isotopes in solution. The decay of the precipitate, as measured with a β-counter, is shown by curve A. Curve B shows the corresponding growth of the four nuclides from ^{228}Ra and ^{224}Ra as measured with the same β-counter. It was obtained by evaporating the filtrate from the precipitate of thorium hydroxide and igniting to drive off ammonium salts. The subsequent fall in curve B is due to the decay of ^{224}Ra when ^{212}Pb has reached transient equilibrium (p. 70). The corresponding rise in curve A is due to the growth of ^{224}Ra from ^{238}Th, which in turn produces ^{212}Pb (see Figure 1.2). The curves were plotted by the author (see Experiment 2).

The rate of decay was however, found to be quite independent of temperature and pressure and therefore could not be a normal chemical process.

1.5. *The disintegration hypothesis*

Shortly after the discovery of radium it was noticed that a radioactive gas was given off from thorium and radium. In the case of thorium the radioactivity of the gas rapidly disappeared, but in the case of radium it persisted for some days. The two gases were called 'emanations'. Both emanations decayed by emission of α-particles. It seemed probable that the α-particle was in fact a positively charged helium atom, since the rate of production of helium by radium agreed approximately with the rate of emission of α-particles. In 1909 Rutherford and T. Royds devised an experiment which gave direct proof of the identity of the α-particle [5] (see Appendix 5, p. 249). This enabled Rutherford to put forward the atomic disintegration hypothesis. Rutherford's work was confirmed by the chemists. Marie Curie had already determined the relative atomic mass of radium as 226·5 and Ramsey and Whytlaw-Gray, by a remarkable piece of work, determined the relative atomic mass of the emanation from radium and showed it to be 222, i.e. approximately 4 units less than radium, the difference being the mass of the helium atom. The gas was called niton, although it is now generally known as radon. It occupied a vacant place, No. 86, in the periodic table and was the last of the noble gases to be discovered.

1.6. *The radioactive series*

Rutherford showed that two series of radioactive elements were formed, one from uranium and one from thorium: later, a third series was discovered called the actinium series (Figures 1.2, 1.3 and 1.4). The stable product of all three series was lead, and lead is found to occur in ores containing uranium, thorium and actinium. The parents of the series are ^{238}U, ^{232}Th, and ^{235}U, respectively. All three of these nuclides decay very slowly, decreasing to half their activity in 10^8 to 10^{10} y. This period is called the half-life (see § 5.3., p. 68). In the process they produce other elements which also decay. The processes by which these elements are formed involve the emission of either an α-particle with a mass of 4 and charge of +2, or a β-particle of negligible mass and charge −1. When an α-particle is emitted, the daughter element produced is two places lower in the periodic table than the parent, whereas when a β-particle is emitted the daughter is one place higher. This is known as the Group Displacement Law. The successive emission of one α- and two β-particles brings an element back to its original position in the table, though its mass number is now four units less.

To summarize these processes: if A stands for the mass number and Z for the atomic number, then α-emission decreases A by four units and decreases Z by two units. The emission of a β-particle, on the other hand, increases Z by one unit with no change in A.

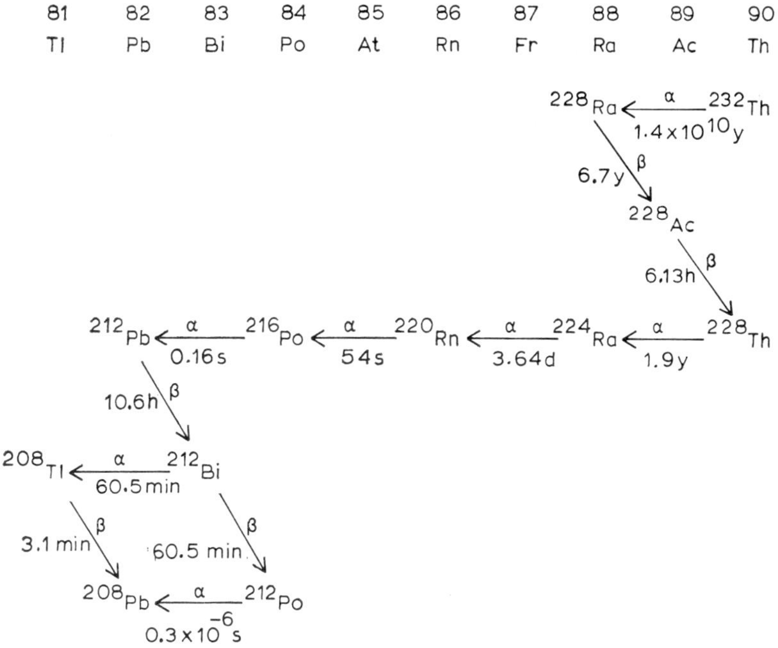

81 82 83 84 85 86 87 88 89 90
Tl Pb Bi Po At Rn Fr Ra Ac Th
$^{228}Ra \xleftarrow[1.4 \times 10^{10} y]{\alpha} {}^{232}Th$
$^{228}Ra \xrightarrow[6.7 y]{\beta} {}^{228}Ac \xrightarrow[6.13 h]{\beta} {}^{228}Th$
$^{212}Pb \xleftarrow[0.16 s]{\alpha} {}^{216}Po \xleftarrow[54 s]{\alpha} {}^{220}Rn \xleftarrow[3.64 d]{\alpha} {}^{224}Ra \xleftarrow[1.9 y]{\alpha} {}^{228}Th$
$^{212}Pb \xrightarrow[10.6 h]{\beta} {}^{212}Bi$
$^{208}Tl \xleftarrow[60.5 min]{\alpha} {}^{212}Bi$
$^{208}Tl \xrightarrow[3.1 min]{\beta} {}^{208}Pb$
$^{212}Bi \xrightarrow[60.5 min]{\beta} {}^{212}Po$
$^{208}Pb \xleftarrow[0.3 \times 10^{-6} s]{\alpha} {}^{212}Po$

Figure 1.2. The thorium ($4n$) series. The numbers beneath or to the left of the arrows refer to the half-life of the preceding isotope. The principal products only are shown. (y=years, d=days, h=hours, min=minutes, and s=seconds.)

By 1913 it had become clear that elements could be chemically indistinguishable, occupying the same place in the periodic table, but that their mass numbers and nuclear properties could be different. In that year Frederick Soddy proposed the name *Isotopes* for such elements, meaning *the same place*.

Each radioactive series can be represented by a simple formula for the mass number (A), in which n is an integer. Thus:

Thorium series	$A = 4n$
Uranium series	$A = 4n + 2$
Actinium series	$A = 4n + 3$

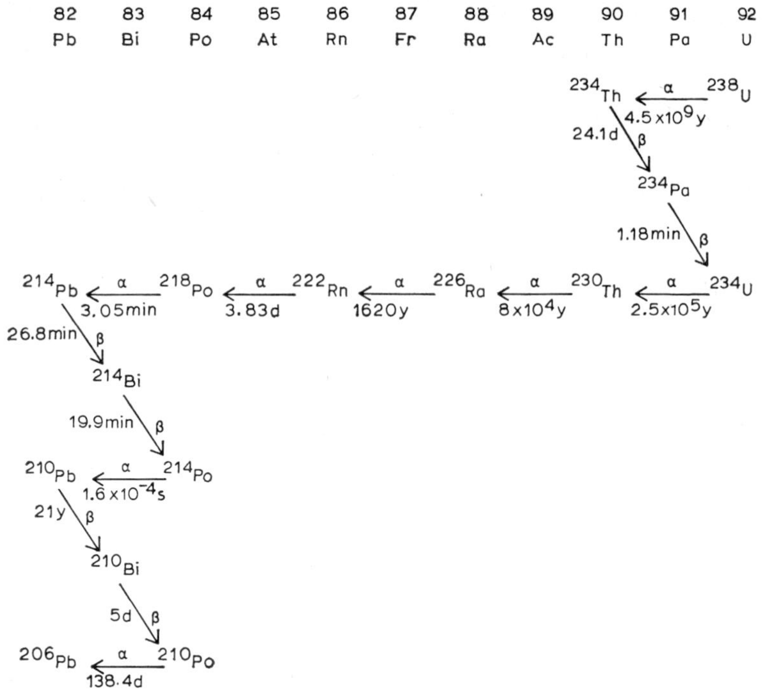

82 Pb | 83 Bi | 84 Po | 85 At | 86 Rn | 87 Fr | 88 Ra | 89 Ac | 90 Th | 91 Pa | 92 U

$^{238}U \xrightarrow[4.5\times10^{9}y]{\alpha} {}^{234}Th \xrightarrow[24.1d]{\beta} {}^{234}Pa \xrightarrow[1.18min]{\beta} {}^{234}U \xrightarrow[2.5\times10^{5}y]{\alpha} {}^{230}Th \xrightarrow[8\times10^{4}y]{\alpha} {}^{226}Ra \xrightarrow[1620y]{\alpha} {}^{222}Rn \xrightarrow[3.83d]{\alpha} {}^{218}Po \xrightarrow[3.05min]{\alpha} {}^{214}Pb \xrightarrow[26.8min]{\beta} {}^{214}Bi \xrightarrow[19.9min]{\beta} {}^{214}Po \xrightarrow[1.6\times10^{-4}s]{\alpha} {}^{210}Pb \xrightarrow[21y]{\beta} {}^{210}Bi \xrightarrow[5d]{\beta} {}^{210}Po \xrightarrow[138.4d]{\alpha} {}^{206}Pb$

Figure 1.3. The uranium ($4n+2$) series. The principal products only are shown.

81 Tl | 82 Pb | 83 Bi | 84 Po | 85 At | 86 Rn | 87 Fr | 88 Ra | 89 Ac | 90 Th | 91 Pa | 92 U

$^{235}U \xrightarrow[7.1\times10^{8}y]{\alpha} {}^{231}Th \xrightarrow[25.6h]{\beta} {}^{231}Pa \xrightarrow[3.4\times10^{4}y]{\alpha} {}^{227}Ac \xrightarrow[22y]{\beta} {}^{227}Th \xrightarrow[18.2d]{\alpha} {}^{223}Ra \xrightarrow[11.7d]{\alpha} {}^{219}Rn \xrightarrow[3.92s]{\alpha} {}^{215}Po \xrightarrow[1.8\times10^{-3}s]{\alpha} {}^{211}Pb \xrightarrow[36min]{\beta} {}^{211}Bi \xrightarrow[2.16min]{\alpha} {}^{207}Tl \xrightarrow[4.79min]{\beta} {}^{207}Pb$

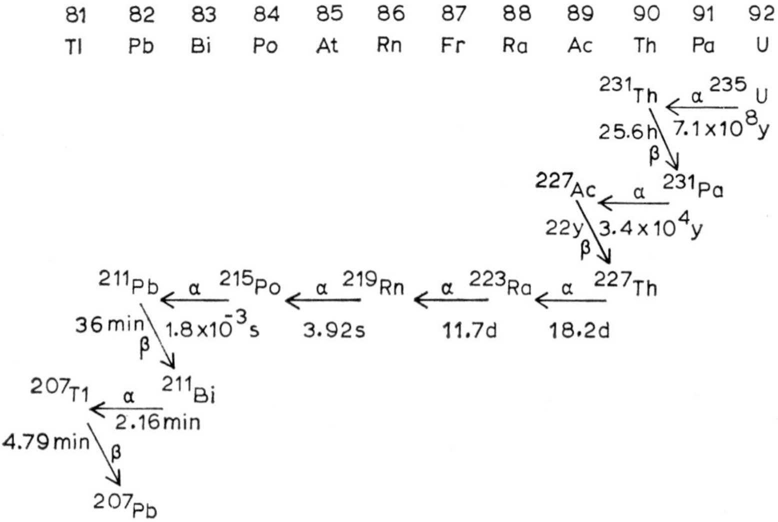

Figure 1.4. The actinium, or ^{235}U ($4n+3$) series. The principal products only are shown.

No $(4n+1)$ series exists naturally: its members are man-made. The parent of the series is neptunium ^{237}Np which has a half-life of 2×10^6y, but only 1 part in 10^{3000} remains of that present at the formation of our galaxy (Figure 1.5).

Each series decays to a different, but stable, isotope of lead. In the thorium series, six α-particles are emitted, resulting in $232-(6\times4)={}^{208}Pb$; the uranium series emits eight α-particles, giving $238-(8\times4)={}^{206}Pb$; and the actinium series emits seven α-particles, giving $235-(7\times4)={}^{207}Pb$.

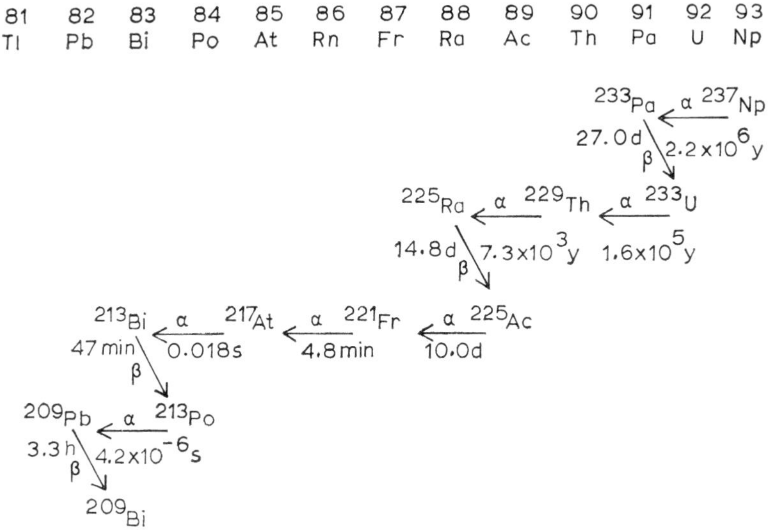

81 Tl 82 Pb 83 Bi 84 Po 85 At 86 Rn 87 Fr 88 Ra 89 Ac 90 Th 91 Pa 92 U 93 Np

$^{237}Np \xrightarrow{\alpha,\ 2.2\times10^6 y} {}^{233}Pa \xrightarrow{\beta,\ 27.0 d} {}^{233}U \xrightarrow{\alpha,\ 1.6\times10^5 y} {}^{229}Th \xrightarrow{\alpha,\ 7.3\times10^3 y} {}^{225}Ra \xrightarrow{\beta,\ 14.8 d} {}^{225}Ac \xrightarrow{\alpha,\ 10.0 d} {}^{221}Fr \xrightarrow{\alpha,\ 4.8 min} {}^{217}At \xrightarrow{\alpha,\ 0.018 s} {}^{213}Bi \xrightarrow{\beta,\ 47 min} {}^{213}Po \xrightarrow{\alpha,\ 4.2\times10^{-6} s} {}^{209}Pb \xrightarrow{\beta,\ 3.3 h} {}^{209}Bi$

Figure 1.5. The artificial $(4n+1)$ series. The principal products only are shown.

To establish the truth of the disintegration theory it was necessary to show that the relative atomic mass of lead obtained from uranium and thorium ores differed from the relative atomic mass of lead obtained from non-radioactive sources. Hönigschmid in Vienna determined the relative atomic mass of lead from pitchblende to be 206·405. Soddy in Oxford obtained 207·694 for lead from a thorium ore. The value for lead from non-radioactive ores is 207·22. The discovery that the relative atomic mass of lead varied according to its source gave strong support to the disintegration hypothesis and confirmed the theory of isotopes.

1.7. *Detection and measurement of the radiations*

The early workers used the blackening of photographic plates and the discharge of electroscopes. The quartz string electrometer coupled to a form of ionisation chamber was used for direct counting, but it has a comparatively long recovery time and could only be used for weak sources. Rutherford developed the zinc sulphide screen for counting individual α-particles by the brief flashes of light (scintillations) produced on impact. He used a low-power microscope to count the scintillations. β-particles also produce a luminous glow, but the glow is far more persistent than with α-particles and the scintillations are feeble. Thin sheets of mica or aluminium foil were introduced between the source and the screen, and the thickness of absorber required to stop the scintillations was determined. This thickness can be expressed in terms of centimetres of dry air at STP, which is more convenient. In this way the kinetic energies of the particles were compared [7].

1.8. *The cloud chamber*

In 1910 C. T. R. Wilson invented his cloud chamber, by means of which the tracks produced by individual ionizing particles could be observed. With the aid of this very beautiful instrument he was able to obtain a series of photographs which has never been surpassed, and which showed by forked tracks the direct collision of atomic nuclei. (For details see Appendix 6, p. 251.) [6].

1.9. *Summary of the results obtained by* 1914

(1) Radioactivity is a random process, i.e. a long lived source will have sensibly constant activity but does not give out exactly the same number of particles per second. This number varies according to a Gaussian distribution (Experiment 1) [8].

(2) The decay is exponential and being quite unaffected by changes of temperature and pressure cannot be a normal chemical process.

(3) The kinetic energies of the particles emitted are very much greater than the mean kinetic energy of gas molecules ($\frac{3}{2}kT$) at ordinary temperature.

(4) Series of radioactive elements are produced by the decay of uranium, thorium and actinium, the end product in each case being lead.

(5) Each radioactive element exists in different forms, the principal differences being in relative atomic mass and half-life, i.e. the time taken for the activity to fall to half its initial value. These different forms of the same element are called isotopes. (Since 1914 other differences have been found, for example in spectral

lines and magnetic properties.) See Figures 1.2–1.5 on pp. 7–9.

(6) The atom consists of a positively charged nucleus where the mass is concentrated, surrounded by negatively charged electrons.

1.10. *The transmutation of nitrogen*

Even before World War I was over, Rutherford resumed his researches. He and Geiger had already shown, in 1908, that α-particles were scattered by atomic nuclei and it seemed possible that the energy of these α-particles could be used to transmute or disrupt an atomic nucleus. In his paper published in 1919 [9] Rutherford showed how he had been able to achieve this by bombarding nitrogen gas with α-particles from radium C(^{214}Bi), which, by β-particle emission, generates ^{214}Po, the source of the α-particles.

This important result was obtained with the simplest type of apparatus—typical of Rutherford. A metal box was fitted with two stopcocks by means of which gases could be introduced. An intense source of radium C(^{214}Bi) was placed inside the box about 3 cm from one end, and an opening in the opposite end was covered with a piece of thin silver foil of stopping power equivalent to about 6 cm of air. The zinc sulphide screen was mounted outside, about 1 mm distant from the silver foil to allow absorbing foils to be placed between them. The whole apparatus was placed in a strong magnetic field to deflect the β-particles. The box was evacuated and the range of the α-particles from radium C was determined (in units of cm of air) by inserting absorbing foils in front of the zinc sulphide screen until no scintillations were observed. The box was then filled with hydrogen and scintillations again became visible. These could only have been produced by protons (hydrogen nuclei) struck by the α-particles from the radium C source. These protons would start with a considerably greater velocity than the α-particles and so have a greater range. When the hydrogen was replaced by oxygen no scintillations were observed on the screen. When the oxygen was replaced by nitrogen, scintillations reappeared. These scintillations were caused by protons produced when an α-particle struck a nitrogen nucleus, transmuting it into oxygen and releasing a proton. The range of these energetic protons was shown to be 28 cm in air, considerably greater than the range of the ' knock-on ' protons produced with hydrogen gas. The reaction can be represented by the nuclear equation shown below, in which the upper number refers to the mass number (A) of the nucleus and the lower to the charge (Z).

$$^{4}_{2}\mathrm{He} + ^{14}_{7}\mathrm{N} \rightarrow ^{17}_{8}\mathrm{O} + ^{1}_{1}\mathrm{H} \qquad (1.4)$$

(When writing nuclear equations it is important to see that both the mass numbers and the charge numbers balance on the two sides of the equation.)

This result, the first clear evidence of artificial transmutation, was confirmed with the aid of a cloud chamber photograph (Figure 1.6). Rutherford continued his experiments using the lighter elements, and soon showed that most of them, up to and including potassium, could be transmuted by the aid of α-particle bombardment. Other workers decided to try protons, but at that time there was no means of accelerating protons to an energy high enough to overcome the potential barrier of the nucleus.

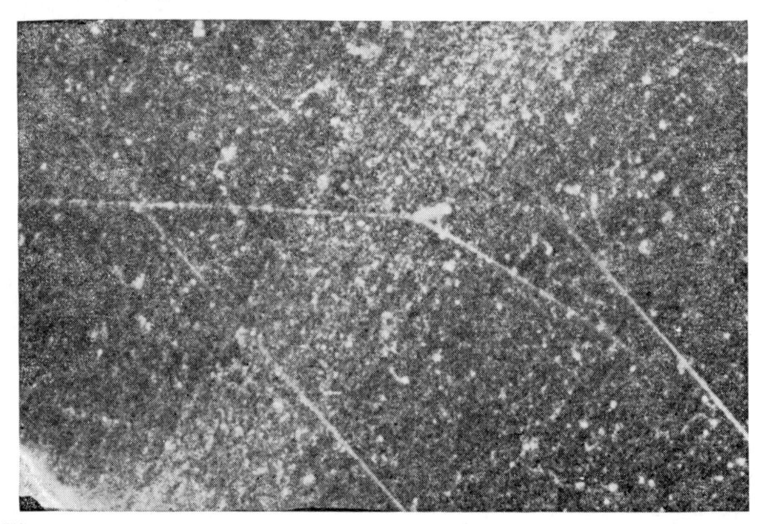

Figure 1.6. A Wilson Cloud Chamber photograph showing the disintegration of the nucleus of a neon atom by a high energy photon. The short broad track is the recoil nucleus (probably an isotope of oxygen) and the two long thin tracks are protons from the disintegration. The reaction is given thus:

$$\mathrm{Ne} + \gamma \rightarrow \mathrm{O} + \mathrm{H} + \mathrm{H} + \text{neutrons}$$

The divergent beam of faint tracks on either side of the star is due to fast electrons ejected from the side wall of the cloud chamber by photons in the incident beam. The transmutation of nitrogen would show a similar forked track. [Photograph by Dr. D. F. Shaw].

1.11. *The splitting of the lithium atom*

In 1932 J. D. Cockcroft and E. T. S. Walton [10] succeeded in accelerating protons to an energy of 300 keV.† They used a lithium

† eV (electron volt) is the energy of an electron after acceleration through a potential of 1 volt.

target. The protons transmuted the lithium to beryllium according to the equation

$$^{7}_{3}Li + ^{1}_{1}H \rightarrow ^{8}_{4}Be \qquad (1.5)$$

and the ^{8}Be nucleus split into two α-particles:

$$^{8}_{4}Be \rightarrow 2\,^{4}_{2}He \qquad (1.6)$$

This was the first instance of nuclear fission. The results were confirmed with a cloud chamber photograph which showed 'hammer' tracks (see below).

The relative masses of the nuclei are Li = 7·0160, H = 1·0073, He = 4·0026. So the mass change in the transmutation of Li to He is: 7·0160 + 1·0073 − 2(4·0026) = 0·0181. This means that 0·0181 g were lost per mole of lithium. According to the Einstein equation $E = mc^2$, where E = energy in J, m = mass in kg and c = velocity of light in m s^{-1} this mass must have been converted into energy, i.e.

$$E = 1{\cdot}81 \times 10^{-5} \times (2{\cdot}99 \times 10^{8})^{2} = 1{\cdot}62 \times 10^{12}\ \mathrm{J\ mol^{-1}}\ \text{of lithium.}$$

This relatively large amount of energy, with the addition of the energy of the incident proton was transferred to the two α-particles. The energy of each α-particle was therefore

$$\frac{1{\cdot}62 \times 10^{12}}{6{\cdot}022 \times 10^{23} \times 2 \times 1{\cdot}602 \times 10^{-19}}\ \mathrm{eV}$$

$$= \left(8{\cdot}40 + \frac{0{\cdot}3}{2}\right)\ \mathrm{MeV}$$

$$= 8{\cdot}55\ \mathrm{MeV} \equiv 1{\cdot}37\ \mathrm{pJ}$$

since the energy of the incident protons was 0·3 MeV.

This value was confirmed by measurements with an ionisation chamber connected to an oscilloscope. Splitting of the beryllium nucleus caused the two α-particles to shoot apart and produced the 'hammer' track in the cloud chamber. 'Hammer tracks' have a long single track formed by the incident particle, and two short tracks in opposite directions, almost at right angles to and at the end of the long track. They resemble the head of a hammer. The two short tracks are produced by the two α-particles flying apart with kinetic energies far in excess of that of the incident proton.

1.12. *The discovery of the neutron*

In 1920 Rutherford had predicted the existence of a neutral particle in the nucleus of the atom, but owing to the extreme difficulty of detection such a particle was not observed for several years. It could not be detected in an ionization chamber, neither would it produce any track in a cloud chamber. In 1930 W. Bothe and H. Becker, while investigating the effect of α-particles on some of the light nuclei, observed that beryllium, on bombardment with α-particles, emitted a highly penetrating type of radiation not previously encountered. This radiation ejected protons from such substances as paraffin wax. The radiation was at first thought to be γ-rays, but in 1932 J. Chadwick [11] suggested that the radiation consisted of neutral particles, i.e. the neutrons predicted by Rutherford, produced as follows:

$$^{9}_{4}\mathrm{Be} + ^{4}_{2}\mathrm{He} \rightarrow ^{12}_{6}\mathrm{C} + ^{1}_{0}\mathrm{n} \tag{1.7}$$

1.13. *Artificial radioactivity*

The discovery of the neutron provided a new nuclear missile for bombarding atomic nuclei and in the hands of Enrico Fermi and Joliot-Curie, son-in-law of Marie Curie, led to the production of a whole range of artificial radioactive isotopes. Their source of neutrons was a mixture of beryllium and polonium, or beryllium and radium. It was soon observed that thermal neutrons, i.e. neutrons with kinetic energies comparable to $\frac{3}{2}kT$, were very much more effective than fast neutrons in causing nuclear reactions. Neutrons could be slowed down by surrounding the source and the target element with paraffin wax or water. Fast neutrons lose energy by collision with protons in the water or paraffin wax. Thermal neutrons can enter the nucleus of the target element, producing an unstable isotope which normally decays by the emission of a β-particle to form a new element one place higher in the periodic table.

1.14. *Nuclear fission* (*a historical survey*)

The discovery of fission is a long and fascinating story covering six years of work by many different scientists.

In 1934, E. Fermi suggested in an article in *Nature* [12] that elements with atomic numbers higher than that of uranium might be produced by bombarding uranium with neutrons. The first such new element was expected to be a homologue of rhenium, as it was assumed that the elements in the seventh period from actinium onwards, were members of a new transition series involving the expansion of the 6d electron shell (see periodic table). This was the first false assumption, but not an unreasonable one (for further discussion see Chapter 4). Fermi made a second suggestion, that isotopes of protoactinium and thorium could also be produced by (n, p) or (n, α) reactions (see Chapter 5, p. 72), or even isotopes of actinium or radium by successive decays.

Fermi irradiated a uranium solution with neutrons from a radium–beryllium source, and separated an activity with a 13-minute half-life by precipitating it by adsorption on manganese(IV) oxide (a process known as co-precipitation). The activity did not appear to resemble any element from lead to uranium and he suggested that it might be eka-rhenium, element number 93. W. Noddack pointed out that manganese(IV) oxide adsorbs a great variety of elements, and at this early stage she intimated that lighter elements could not be ruled out.

The following year, 1935, O. Hahn and L. Meitner published a paper confirming the 13-min activity and also claimed the detection of a 90-min activity. They proved that these two activities did not resemble uranium, protoactinium or thorium. However, further work by the same researchers published in 1936 and 1937 failed to confirm the existence of the 13-min and 90-min activities, but they made the important observation of a 23-minute activity, which they correctly assigned to uranium. (This is ^{239}U, which decays as Fermi predicted, by β-emission to ^{239}Np ($t_{1/2}$=2·3 d) which is the parent of ^{239}Pu). Hahn and Meitner failed to detect the ^{239}Np, which is not in the least surprising. Further work by Hahn, Meitner and F. Strassmann detected other activities, which appeared to resemble osmium, iridium and platinum, and these were thought to be elements numbers 94, 95 and 96. An article published in the Chemical Review for 1938 [13] discussed the evidence for the existence of transuranic elements in some detail. It is now known that the conclusions were nearly all incorrect. However, the author of the article did point out a number of inconsistencies which helped to lead to the correct interpretation the following year.

In the meanwhile other workers followed up Fermi's second suggestion, detecting activities which seemed to resemble actinium and radium chemically. These observations led J. Curie and P. Savitch into the field, and in 1938 [14] they detected a 3·5 h activity which was supposed to be an actinium isotope. To confirm this they mixed the 'activity' with actinium-228, the first decay product of natural thorium, and lanthanum. On separation they found that the 3·5 h activity followed lanthanum and not actinium. In this they were very near to discovering the truth. Unfortunately, they found that their activity could be partially separated from lanthanum, and this must have misled them. The 3·5 h activity was probably a mixture of lanthanum and yttrium. Both of these elements are known to have isotopes with half-lives close to 3·5 h, (^{141}La and ^{92}Y).

In 1938 Hahn and Strassmann (Meitner had fled to Sweden from the Nazi Government) repeated Curie and Savitch's work, but concentrated on the supposed radium isotopes. Several of these had been recognized as products of neutron bombardment of uranium, and all produced daughter products. Hahn and Strassmann examined the alleged Ra-IV, which had a half-life of 10–13 d. They mixed it with ^{224}Ra (obtained from natural thorium) and added barium as a carrier. They separated the radium and barium by fractional crystallization of the bromides, as in Marie Curie's original work, and found that the 10–13 day activity followed the barium. They also investigated the 3·5 h alleged actinium isotope, and confirmed Curie and Savitch's work that the 3·5 h activity followed lanthanum and not actinium. This led them to publish, in 1939, their famous paper *The detection and behaviour of alkaline-earth metals formed in the irradiation of uranium with neutrons* [15] in which they put forward the theory of fission. In a second paper they showed that 'Ra III' with $t_{1/2}=86$ min was identical with barium-139 ($t_{1/2}=85$ min) formed by the neutron irradiation of natural barium, and 'Ac IV' ($t_{1/2}=35$ h) was identical with lanthanum-140 ($t_{1/2}=40$ h) formed by the neutron irradiation of natural lanthanum. The lighter elements, strontium and yttrium, were also recognized among the products of neutron irradiation of uranium. By blowing air through neutron irradiated solutions of uranyl(VI) nitrate, krypton and xenon isotopes were removed.

Following the publication of these papers, Meitner and O. R. Frisch [16] realized that the fission process would release large amounts of energy and was therefore thermodynamically feasible. Their ideas were confirmed by observation of the tracks in a cloud chamber and the large pulses observed from an ionization chamber. The energy liberated was found to be about 200 MeV (32 pJ) per atom of uranium, and several neutrons were simultaneously released. It was immediately realized that these neutrons would be capable of producing a self-sustained chain reaction with the release of enormous amounts of energy.

Thus in the period 1920–1939 very great advances had been made. These can be summarized as follows:

(1) The artificial transmutation of nitrogen by Rutherford.

(2) The splitting of the nucleus by Cockroft and Walton.

(3) The discovery of the neutron by Chadwick.

(4) The production of a whole range of artificial radioactive elements by neutron bombardment of stable elements.

(5) Perhaps the biggest discovery of all, the fission of uranium.

Questions

1.1. What do you understand by the statement that 'The half-life of the radium isotope ^{226}Ra is 1590 years'? Determine the Avogadro constant, given that $1{\cdot}16 \times 10^{18}$ α-particles were emitted in a radioactive disintegration and over the same time, $0{\cdot}043$ cm^3 of helium was collected at STP. (Oxford Schol.).

1.2. What is meant by the half-life of a radioactive element? Why is the half-life independent of the chemical composition and the mass of the sample?

1.3. What is the disintegration rate (s^{-1}) of:

(*a*) 1 g of uranium oxide (U_3O_8) six months after preparation from a uranium ore;

(*b*) 1 g of Joachimstal pitchblende containing 60% by mass of uranium? (Half-life of $^{238}U = 4{\cdot}5 \times 10^9$ years; ignore the ^{235}U.)

2. The control and use of nuclear energy

2.1. *Introduction*

Nuclear energy was developed for both peaceful and destructive purposes in six years from the discovery of the fission of uranium. The rapidity was due to the spur of the Second World War, for it was realized that a nuclear bomb would be more powerful than a normal high explosive bomb by at least four orders of magnitude. The use of nuclear energy for the generation of electrical energy and the production of artificial radioactive isotopes followed as a natural consequence.

With the advent of war, a blanket of secrecy soon descended on the results of further experiments concerned with fission. The position in 1939 was recorded in the Annual Reports of the Chemical Society and the possibility of producing a super-bomb was discussed. At that time it was thought that it would require a sphere of pure uranium 3 m in diameter weighing 40 tonnes and, as no known aircraft could possibly carry such a bomb, the use of uranium fission in this field seemed unlikely. Nothing further was published until 1945.

By June 1940 the following information was known:

(1) The nuclei of uranium, thorium and protactinium could all undergo fission by neutrons.

(2) Both fast and thermal neutrons would cause fission of uranium, but ^{235}U alone was fissile to thermal neutrons.

(3) Thermal neutrons could be captured by ^{238}U. Neutrons of particular energies are also captured, a process called resonance capture. These resonances occur in the energy band between 6 and 200 eV (0·96–32 aJ) (Figure 2.2).

(4) The average number of neutrons released per fission was between 2 and 3.

The problem of producing a self-sustaining chain reaction was complicated due to the capture of neutrons by ^{238}U. This neutron capture was further investigated and late in 1940 E. McMillan and P. Abelson discovered element 93 [17]. At that time the possibility of forming transuranic elements had been largely discredited. McMillan and Abelson bombarded thin uranium foil with neutrons and the fission products escaped from the foil by recoil. They observed two activities remaining in the foil, one with a half-life of 23 min and the other with a half-life of 2·3 d. The 23-min activity could not be separated from uranium and was identified as ^{239}U—which had been discovered by Hahn and Meitner in 1937 (p. 15).

The 23 min and 2·3 d activities both emitted β-particles. The 2·3 d activity could be separated by use of a reduction–oxidation cycle and showed little resemblance to rare earth elements. It was identified as a definite transuranic element and given the name of neptunium (Np). The following year the isotope of plutonium, ^{239}Pu, was discovered as a decay product of ^{239}Np. [18]. From theoretical considerations it seemed that ^{239}Pu would be fissile like ^{235}U. It decayed by α-particle emission with a long half-life. Fission is most likely with heavy nuclei of even Z, but odd A. The problem of producing a bomb resolved itself into two possibilities: to separate ^{235}U from ^{238}U, or to produce ^{239}Pu. As several kilograms of each would be required the task was formidable.

2.2. *Isotope separation*

The separation of isotopes on a significant scale was at that time in its infancy. H. C. Urey had separated deuterium from hydrogen by the fractional electrolysis of water [19], but this was a simple problem compared to the separation of ^{235}U and ^{238}U. Uranium forms one volatile compound: uranium hexafluoride. This has a vapour pressure of one atmosphere at 60°C, but is intensely corrosive.

The two principal methods considered for the separation of the uranium isotopes were:

(1) Gaseous diffusion.

(2) Electromagnetic separation.

2.2.1. *Gaseous diffusion.* The percentage of ^{235}U in natural uranium is 0·7, i.e. one atom of ^{235}U to every 140 atoms of ^{238}U. To produce 90 per cent ^{235}U this ratio would have to be increased to 9 ^{235}U to 1 ^{238}U. The overall separation factor required was therefore 1260, i.e. 9×140. In any given process there is an ideal separation factor which operates at the start before appreciable separation has occurred. In the case of gaseous diffusion this ideal separation factor is:

$$i=\sqrt{\frac{M_2}{M_1}}$$

where M_2 and M_1 are the relative molecular masses of the two substances concerned. As the gas used would have to be UF_6, and natural fluorine has no isotopes:

$$i=\sqrt{\frac{352}{349}}=1{\cdot}0043.$$

As the process of diffusion continues the separation factor becomes smaller. When half of the gas has diffused the separation factor becomes 1·003. It was, therefore, clear that a great many stages would be required (an estimate of 4000 was given) and continual recycling as in fractional distillation. The problem was further complicated by the corrosive nature of UF_6.

To ensure true diffusive flow it is essential that the size of the holes in the diffusion barrier should be about one tenth of the mean free path of the molecules. The mean free path of a molecule of UF_6 is about 10^{-8} cm at atmospheric pressure. It would be desirable to maintain atmospheric pressure on one side of the barrier and a very low pressure on the other. In view of the large number of stages required and the necessity of having a large barrier area in the early stages, it was estimated that acres of barrier would be needed, and this barrier must have uniform holes of 10^{-9} cm diameter. In addition, it must resist corrosion by UF_6 and be able to withstand one atmosphere pressure.

At each stage in the process the diffused gas would have to be pumped to the next stage and the residue, to be mixed with the diffused gas from the previous stage, would also have to be pumped back to one atmosphere pressure. This would necessitate thousands of pumps which also must be corrosion resistant, and the entire system must be vacuum tight. All this involved chemical engineering problems of a high order, and it was a great triumph when the process was finally operating.

2.2.2. *Electromagnetic separation.* Owing to the engineering difficulties of gaseous diffusion it was decided to attempt, at the same time, electromagnetic separation. This involved applying the principle of a mass spectrometer to a preparative process. There are considerable difficulties in doing this. A mass spectrometer is designed to give a very sharp separation, but in an extremely small yield. It can normally handle quantities of the order of a fraction of a microgram per hour. One of the main factors in limiting the output of a mass spectrometer is the difficulty of producing large quantities of gaseous ions.

Late in 1941 E. O. Lawrence, the inventor of the cyclotron, was put in charge of this work and set up a large mass spectrometer using the principle of Dempster's mass spectrograph (Figure 2.1). He used as a magnet his 0·94 m cyclotron magnet. By the end of the year he had effected a high degree of separation of the isotopes but in small yield; his separation factor was, however, very much greater than that obtained by a single stage of gaseous diffusion.

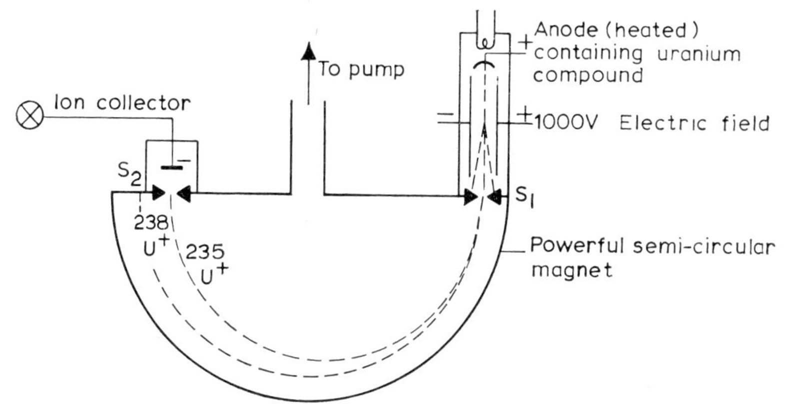

Figure 2.1. The electromagnetic separation of ^{235}U from ^{238}U. E. O. Lawrence applied the principle of Dempster's mass spectrometer. The ions of ^{235}U and ^{238}U, principally U^+, were accelerated by the electric field and passed through the slit S_1 into the evacuated chamber of the magnetic field. Ions with the same value of e/m, even if they followed slightly divergent paths, were brought to the same focus by the semi-circular magnetic field. A giant magnet was used, with a pole diameter of 4·67 m.

To produce a separation on the necessary scale it was essential to use a much larger magnet, and so increase the angular separation of particles of different masses. One, in fact, was already under construction for a giant cyclotron, and this magnet was called into use. The pole diameter was 4·67 m and this produced a much improved degree of separation. Other methods of separation were tried, but the electromagnetic method was the first to separate significant quantities of ^{235}U. ^{235}U was in fact used in the production of the first nuclear bomb.

2.3. *The chain reaction*

To produce plutonium in quantity it was necessary to carry out a chain reaction in uranium, but at the same time allow some of the fission neutrons to be captured. This was a matter of considerable difficulty. The chief problem of the chain reaction concerned the fate of the neutrons released in fission. They could:

(1) Escape from the mass.
(2) Be captured by impurities.
(3) Cause further fissions.
(4) Be captured by uranium.

Process 1 could be reduced by using a large mass of uranium. Process 2 could be reduced to very small proportions by purifying the uranium but, for the chain reaction to succeed, it would be essential that process 3 should be made more probable than process 4, otherwise the chain reaction could not be self-sustained. The relative rates of processes 3 and 4 are fixed by the relative cross-sections. (The term 'cross-section' is a measure of the probability that an approaching particle will penetrate the nucleus and cause a nuclear reaction.) It may be considered as the effective area of the target nucleus within which the reaction will occur if the projectile strikes it. When a neutron enters a nucleus, two things may happen: either the nucleus may split in two in a matter of microseconds, or the neutron may be retained and the surplus energy given off in the form of γ-radiation. The probability of the first process occurring is called the fission cross-section and the probability of the second occurring the capture cross-section. Cross-sections are measured in 'barns'. One barn (b) is 10^{-28} m^2 and this is the approximate cross-sectional area of an average nucleus. Cross-sections vary considerably. For fast neutrons and charged particles cross-sections are of the order of 1 b or less. For thermal neutrons they vary widely from 10^4 to 10^{-4} b.

The ^{235}U nucleus has a cross-section for fission by thermal neutrons of about 600 b(σ_f) and a cross-section for capture of thermal neutrons of 100 b (σ_a). ^{238}U is not fissionable by thermal neutrons, but has a capture cross-section of 2·3 b.

As stated previously, the proportion of ^{235}U in natural uranium is 0·7%. In the absence of resonance capture by ^{238}U the chance of fission occurring in pure natural uranium may be expressed as:

$$\frac{\text{Fission cross-section} \times \text{percentage of } ^{235}\text{U}}{(\text{Capture cross-section} \times \text{percentage of } ^{238}\text{U}) + (\text{capture cross-section} \times \text{percentage of } ^{235}\text{U})}$$

$$= \frac{600 \times 0{\cdot}7}{(2{\cdot}3 \times 99{\cdot}3) + (100 \times 0{\cdot}7)}$$

$$= 1{\cdot}41$$

This ratio is known as the *reproduction factor*. Resonance capture (p. 18) reduces the reproduction factor to less than 1 so that no chain reaction is possible in pure natural uranium. The reproduction factor can be increased slightly by using a moderator to slow down the neutrons rapidly through the critical region of 6–200 eV (0·96–32 aJ) energy (Figure 2.2 and p. 39) and by the correct spacing of the uranium in the moderator. The main means of alteration is by using uranium enriched in ^{235}U, but this method could not be used in the early stages of development. The moderator must consist of light atoms so that the neutrons lose their energy in a relatively small number of collisions, and it must also have little or no tendency to capture neutrons. Considering the elements in

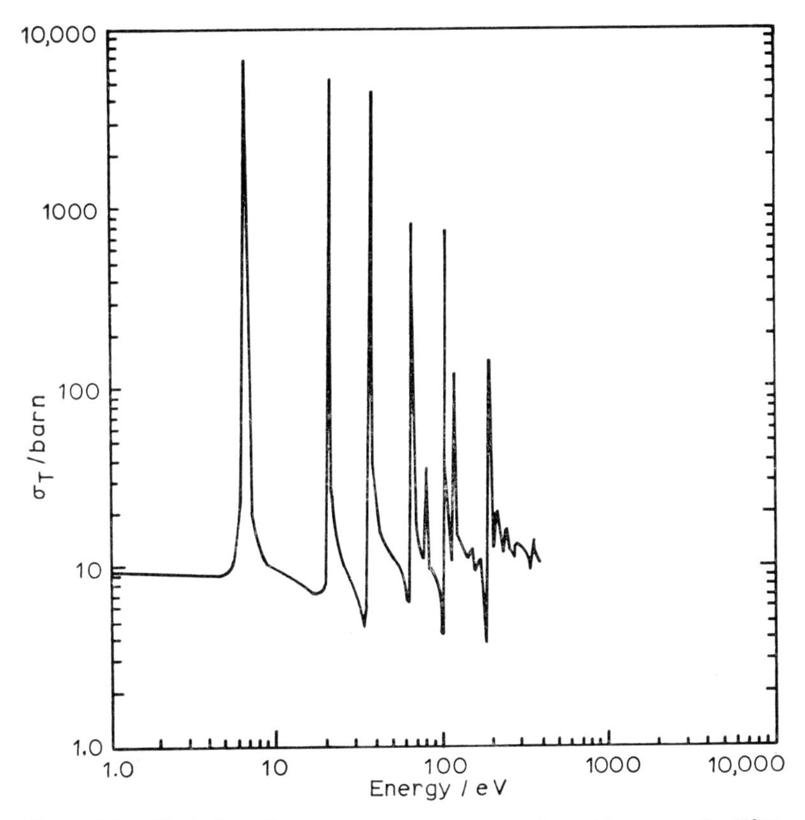

Figure 2.2. Variation of neutron capture cross-section with energy for ^{238}U. Note the very high cross-sections and the narrow energy bands in the region 6–100 eV. These are the resonance capture bands. σ_T is the resonance capture cross-section. (Reproduced, by permission, from *Man Made Transuranium Elements* by G. T. Seaborg, Prentice Hall Inc., Englewood Cliffs, N.Y., 1963.)

order, hydrogen would be the obvious choice, but unfortunately the capture cross-section (0·3 b) for thermal neutrons is too large and it can only be used with uranium enriched in ^{235}U. Deuterium in the form of heavy water is suitable, but heavy water was not available in quantity in 1940. Helium is ideal, but unsuitable, because it cannot readily be obtained in a dense form. The light isotope of lithium has a thermal neutron cross-section, σ_a, of 926 b. Beryllium is quite suitable, but very scarce, and the preparation of pure beryllium is difficult. The light isotope of boron has a very high capture cross-section (4000 b) and so this leaves carbon, which has a cross-section, σ_a, of 0·004 b. Very pure graphite was chosen as the moderator for the first pile to be built.

As regards the spacing of the uranium in the moderator, Fermi and L. Szilard suggested arranging it in lumps, or aggregates, rather than dispersing it evenly. The advantage of this design was mainly convenience of construction, but it also reduced the proportion of neutrons captured by resonance absorption. The resonance-capture cross-sections are so large that this process takes place mainly in the surface of the lumps, i.e. is proportional to r^2 where r is the radius of the lump. The capture of thermal neutrons, however, is approximately proportional to the total volume, i.e. to r^3. It is therefore advantageous to have a relatively high volume to surface ratio for the uranium metal.

The diffusion length for fast neutrons is many centimetres and therefore it was not necessary to make a homogeneous assembly. The high energy neutrons emitted in fission from one lump would be reduced to thermal energies in the moderator before entering another lump. It was assumed that only those neutrons which were reduced to the resonance absorption energy inside a lump of uranium or were turned back by the moderator before they reached thermal energies would be captured. (This assumption proved to be an oversimplification for it was subsequently discovered (p. 19) that ^{238}U has a capture cross-section for thermal neutrons of 2·3 b.) The size of the lumps and the correct spacing in the moderator determined the overall size of the assembly, and possibly the success or failure of the project. As all the necessary parameters, particularly the resonance absorption energies, were not accurately known, an experimental approach was essential.

2.4. *Multiplication factor*

In order to produce a chain reaction it is essential that at least one neutron produced in a fission is able to cause a further fission. The ratio of the number of neutrons generated to the number lost is called the multiplication factor, k. The following quantities determine the value of k:

(1) The number of neutrons produced per fission (N).

(2) The probability (p) that a given neutron will escape absorption by ^{238}U and ^{235}U—this will be <1.

(3) The probability (f) that a given neutron will be absorbed by ^{235}U and produce a further fission—also <1.

Then

$$k_{\infty} = Npf \tag{2.1}$$

k_{∞} is the multiplication factor in a pile of infinite size, i.e. one in which no neutrons escape from the mass. In a pile of finite size k will be smaller than k_{∞}, but the loss of neutrons can be reduced by surrounding the pile by a neutron reflector such as graphite or beryllium. However, there is a certain minimum size below which k cannot be greater than one, since the number of neutrons produced per second is proportional to r^3 whereas the number lost through the surface is proportional to r^2. This size is called the critical size and it depends on the moderator used (see p. 39 for further details).

2.5. *The uranium pile*

In the middle of 1941 the first experimental pile was set up. Fermi designed this pile, which consisted of a graphite cube of size 2·4 m and contained 7 tonnes of uranium oxide, arranged in lumps clad in iron, and spaced at equal intervals throughout the graphite. The object of this set-up was to determine the size of the pile which would give a k factor >1. The whole success or failure of the operation depended upon k. If k were greater than 1 the chain reaction would proceed. If k were less than 1 it would not proceed. It would be essential to design a pile with a k factor of at least 1·02 to allow for possible capture of neutrons by fission products which could reduce k to <1 and so halt the reaction. Experiments were conducted by using a neutron source outside the pile and determining the neutron flux at various points in the mass, both in the presence of uranium and in the absence of uranium. This experiment showed that the neutron flux increased in the presence of uranium, but the pile was obviously too small for a self-sustained chain reaction although it enabled plans to be drawn up for the construction of a much larger pile.

There was, however, one further serious problem and that was the danger of the chain reaction getting out of control. Once $k>1$

the reaction accelerates very rapidly. The rate of increase is determined by τ, the average time between successive generations of neutrons. If N is the number of neutrons present at any one time, then

$$\frac{dN}{dt} = \frac{N(k-1)}{\tau}$$

since $(k-1)$ is the fractional increase in the number of neutrons. Integrating this equation with respect to t gives:

$$\ln N = \ln N_0 + \frac{(k-1)t}{\tau}$$

or

$$N = N_0 \exp [(k-1)t/\tau] \quad (2.2)$$

N_0 being the number of neutrons when $t=0$.

The value of τ, the generation time in a graphite moderated reactor, is about 10^{-3} s. Suppose k were made 1·01. Then from Equation (2.2)

$$N = N_0 \exp (10t)$$

so that the number of neutrons in the pile will be multiplied by e, i.e. 2·71 every 0·1 s. In one second the number of neutrons would increase by a factor of 20 000 and the whole mass would become incandescent. The heat could be removed by cooling, but this does not enable the chain reaction to be controlled. To control the chain reaction it is necessary to insert absorbers in the form of boron or cadmium rods, which have very high capture cross-sections for neutrons. The above calculation shows, however, that it would not be possible to introduce these rods sufficiently rapidly to effect control. However, a very fortunate discovery was made, namely that the emission of some neutrons was delayed (see p. 27) by up to as much as a minute. This gives an inertia to the system and τ is increased to about 10^{-1} s. With the above-mentioned value of k the doubling period for the neutron population becomes about 10 s, which enables the chain reaction to be controlled, and hence permits the construction of a pile.

2.6. *The first self-sustained pile*

In the autumn of 1942 sufficient quantities of pure uranium, uranium oxide and graphite were available to attempt the construction of a pile in which the chain reaction would be self-sustained. The pile was built with the uranium metal in the form of lumps spaced at uniform distances throughout the graphite, and control rods of cadmium were inserted. This was a fortunate precaution as the pile became critical at a smaller size than had been expected. By the end of the year the pile was producing energy at the rate of 200 W.

Since the main object of constructing this pile was to produce plutonium, it was important to determine the rate of plutonium production in relation to the energy released. As about 2·5 neutrons are released per fission, and at least one neutron is needed to keep the pile going, little more than one atom of plutonium could be produced per fission. Each fission involves the release of some 200 MeV (32 pJ) of energy, so a pile working at 200 W would produce:

$$\frac{200}{32 \times 10^{-12}} \text{ fissions s}^{-1}$$

$$\text{and } 6{\cdot}25 \times 10^{12} \text{ atoms of plutonium s}^{-1}.$$

To produce 10 kg of plutonium in a year the rate of working would have to be increased by at least a factor of 10^6. This shows the magnitude of the task.

2.7. *The process of fission*

When the ^{235}U nucleus captures a thermal neutron, the ^{236}U isotope is produced in an excited state, i.e. a state in which the nucleus contains excess potential energy, and either decays to the ground state by the emission of one or more γ-rays, or undergoes fission with the release of two or three neutrons. The mass and nuclear charge of the fission products vary considerably. Figure 2.3 shows the yield of fission products plotted against atomic number. It will be noted that two peaks occur around mass numbers 95 and 140.

The uranium nucleus has a considerably higher neutron/proton ratio than the nuclei of the stable isotopes of the elements produced in fission. As a result the fission products are strongly radioactive and the primary products decay by a chain such as:

$$^{131}_{50}\text{Sn} \xrightarrow[3{\cdot}4\,\text{m}]{\beta} {}^{131}_{51}\text{Sb} \xrightarrow[23\,\text{m}]{\beta} {}^{131}_{52}\text{Te} \xrightarrow[25\,\text{m}]{\beta} {}^{131}_{53}\text{I} \xrightarrow[8\,\text{d}]{\beta} {}^{131}_{54}\text{Xe}$$

with the emission of β-particles. One or two, notably ^{87}Br and ^{137}I, decay by emitting neutrons as well as β-particles. This type of decay is rare and only occurs when the nuclide has an exceptionally high neutron emission probability. The neutrons emitted by these isotopes are called 'delayed' neutrons as opposed to the prompt neutrons emitted in fission. The half-lives of these two isotopes are 54·5 s and 24 s respectively, and they are principally responsible for the 'inertia' mentioned on p. 26.

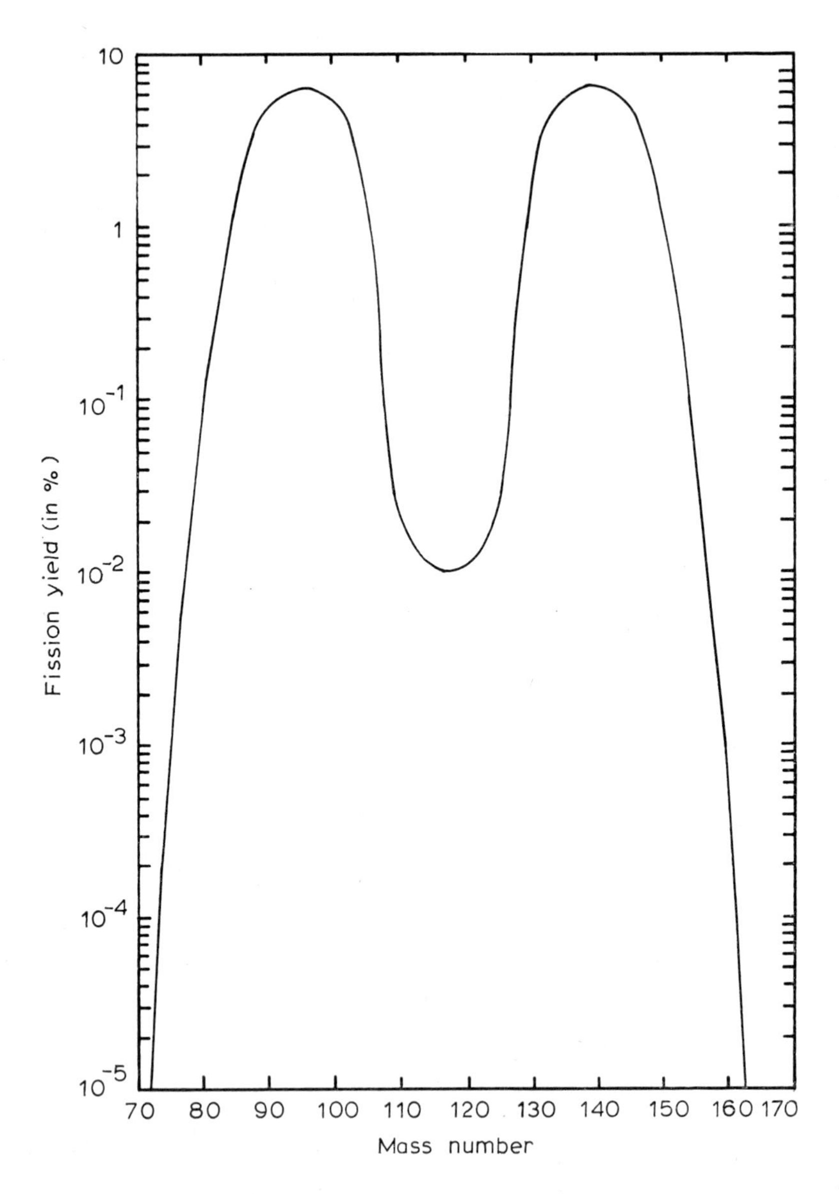

Figure 2.3. Yields of fission-products as a function of mass number for the slow neutron fission of ^{235}U. (Reproduced, with slight modification, from *Nuclear and Radiochemistry* by G. Friedlander, J. W. Kennedy and J. M. Miller (1964) with permission from John Wiley & Sons.)

2.8. *Chemical separation*

While the physicists were studying the production of plutonium, the chemists were studying its chemistry. ^{239}Pu was produced by bombardment of a large quantity of uranium with neutrons produced from a cyclotron. The quantity obtained was 0·5 mg, more than enough to carry out tracer studies. The experiments showed that plutonium did not resemble osmium as was at first expected, but showed considerable resemblance to uranium with oxidation states of 3, 4, 5 and 6. This was really the first evidence for the existence of the actinide or second rare-earth series.

When uranium rods are withdrawn from a pile it is necessary to allow the short-lived fission products to decay, but after the uranium rods have 'cooled' for ten days there are a large number of highly active elements present. These elements, the products of fission, can be divided into two groups: the heavy group with mass numbers from 127 to 154 and the light group with mass numbers from 83 to 115. These represent some thirty different elements, but some of these have very short half-lives. However, at least fifteen are present in significant concentration. They all eventually decay to stable isotopes of the natural elements.

The first method of separation used lanthanum fluoride as carrier in an oxidation–reduction cycle. The principle of the method depends on the fact that plutonium can be reduced to Pu(IV) by relatively mild reducing agents which leave the uranium as U(VI). As Pu(IV), it can be precipitated as a fluoride with lanthanum fluoride as a carrier, leaving the U(VI) in solution. Unfortunately, fission products are also carried by lanthanum fluoride, so that a preliminary precipitation of these is necessary when the plutonium is in solution as Pu(VI).

An improved method of separation was developed by G. T. Seaborg and S. G. Thompson using bismuth phosphate [20], which is a specific carrier for Pu(IV) but only carries a small amount of fission products. Bismuth phosphate has the additional advantage that it will precipitate in 2 M nitric acid. A complex series of oxidation–reduction cycles yielded pure PuO_2, but, in the first instance, only 1 μg was obtained. So urgent was the problem that the whole process had to be scaled up to the gram stage in a single step. It was essential to develop methods of remote control so that the whole process could be carried out in concrete compartments below ground until the highly active fission products had been removed. The pure plutonium compounds produced had to be handled with extreme care. Plutonium, owing to its α-activity and chemical properties, is one of the most dangerous substances known.

The maximum permissible quantity in the body of an occupational worker is 0·2 μg.

2.9. *The production of plutonium*

The small experimental pile built in 1942 by Fermi had demonstrated that a chain reaction was possible. It was therefore decided early in 1943 to construct a larger pile as a pilot plant, so that gram quantities of plutonium could be obtained for experimental purposes, and to discover some of the difficulties which might be encountered in producing a pile to yield kilogram quantities of plutonium. In this pilot pile the uranium, clad in aluminium, was inserted in the form of rods so that it could be easily withdrawn. These rods were situated in channels in the graphite moderator so that air could be passed over them as a coolant. The cooling air became active due to neutron activation of argon. ^{40}Ar yields ^{41}Ar which decays by β-particle emission with a half-life of 109 min. The air had to be discharged through a tall chimney. This pile was designed to operate at 1 MW, ten thousand times the power level of the first pile, and by the end of the year it was successfully producing gram quantities of plutonium. Much larger piles were then built to run at 300 MW and thus produce kilogram quantities of plutonium. Work on these larger piles was started in the middle of 1943 before the 'pilot' pile was in operation. In the event it was decided to use water cooling rather than air cooling. The first of the large piles was working by the autumn of 1944. Two more were constructed and the whole plant was operating by the summer of 1945.

2.10. *The fission bomb*

The operation of the pile was essentially a slow process involving the controlled release of the energy of fission. For maximum explosive effect it would be necessary to release the energy in the shortest possible time. The slow neutron reaction in the pile would be of no use for this purpose since the effective life, or generation time τ, for thermal neutrons is 10^{-3} s (p. 26) and the parts of the bomb would separate before any major explosion occurred. It was therefore essential to use pure ^{235}U or ^{239}Pu and a fast neutron reaction. Though resonance capture does occur with both these nuclei the probability is much smaller than with ^{238}U. An explosion will occur provided the number of neutrons which escape is less than the number which cause further fissions, and this is determined by the critical mass. The problem, therefore, was to bring two subcritical masses together and hold them in contact long enough to produce an explosion.

2.11. *Critical mass*

An estimate of critical mass can be obtained by assuming that the volume will be the volume of a sphere of radius equal to the mean free path of the neutrons between fission reactions. Let this radius be r. The mean free path will be equal to $1/n\sigma$ where n is the number of nuclei per m^3 and σ is the capture cross-section (p. 22).

$$n = \frac{6{\cdot}02 \times 10^{23}}{235 \times 10^{-3}} \times \rho \text{ where } \rho \text{ is the density of uranium}$$
$(= 19\,000\ \text{kg m}^{-3})$.

σ for a fast neutron reaction can be taken as equal to the cross-sectional area of a nucleus which for $^{235}U \approx 2{\cdot}5 \times 10^{-28}\ m^2$.

$$\text{Hence: } r = \frac{1}{n\sigma} = \frac{235 \times 10^{-3}}{6{\cdot}02 \times 10^{23} \times 19\,000 \times 2{\cdot}5 \times 10^{-28}}\ \text{m}$$
$$= 0{\cdot}0822\ \text{m},$$

and the critical mass, $m = \frac{4}{3}\pi r^3 \rho$

$$= \frac{4\pi \times (0{\cdot}0822)^3 \times 19\,000}{3}$$
$$\approx 44\ \text{kg}.$$

This is an oversimplified and somewhat crude calculation, but it gives approximately the correct answer for ^{235}U. The critical mass can be reduced by surrounding the fissile material with a neutron reflector or tamper. Taking these two considerations into account the critical mass was estimated to be about 30 kg.

It is instructive to calculate the time required for the fission of all the atoms in 80 kg of ^{235}U (see question 2.3). To reduce this time to a minimum it is essential to make k, the reproduction factor, much larger than in a pile, say about 2, and this requires that the final mass must be nearly equal to two critical masses. When the critical mass is exceeded a chain reaction cannot be prevented. There are always stray neutrons from cosmic rays, and spontaneous fissions occur. The problem which had to be solved was how to bring two sub-critical masses together and to hold them in contact long enough for the fission of a considerable portion of the whole. Failure to achieve a sufficiently rapid assembly would result in a premature explosion which would merely separate the parts of the bomb. The problem was solved by the use of a tamper of dense material which serves both to reflect neutrons back into the core and to increase the inertia, and by assembling the two sub-critical masses by shooting one as a projectile in a gun against the other as a target.

2.12. *The fusion bomb*

Figure 2.4 shows that energy can be released by fusion, i.e. by producing heavier elements from lighter ones, as well as by fission. Both processes reach an equilibrium about iron ($Z=26$). Fusion is a more difficult process to achieve than fission because of the Coulomb force of repulsion. According to Coulomb's law,

$$\text{force of repulsion} = \frac{e_1 e_2}{4\pi\epsilon_0 r^2} \tag{2.3}$$

where e_1 and e_2 are the respective charges on the nuclei, ϵ_0 is the permittivity of free space, and r is the distance between them.

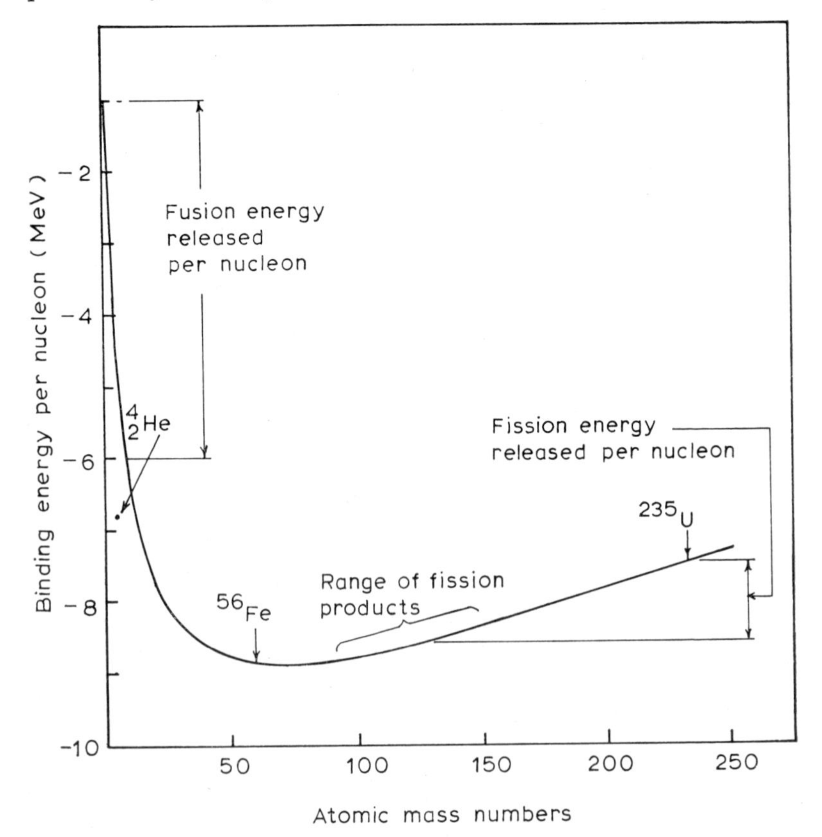

Figure 2.4. The curve of binding energy.

According to this equation the force would increase rapidly as the two nuclei approached and become infinite when they were in contact. If this in fact were so, fusion would be impossible, and no elements with Z greater than 1 could exist. However, there is another force, the nuclear force, which is not yet clearly understood. It operates over very small distances only (about 5×10^{-15} m), is strongly attractive, being many times greater than the Coulomb force at these distances, and is independent of the nature of the nucleon.

Consider a charged particle approaching a nucleus from an infinite distance, with kinetic energy E_0 (Figure 2.5). It loses kinetic energy in overcoming the Coulomb force and gains potential energy. If

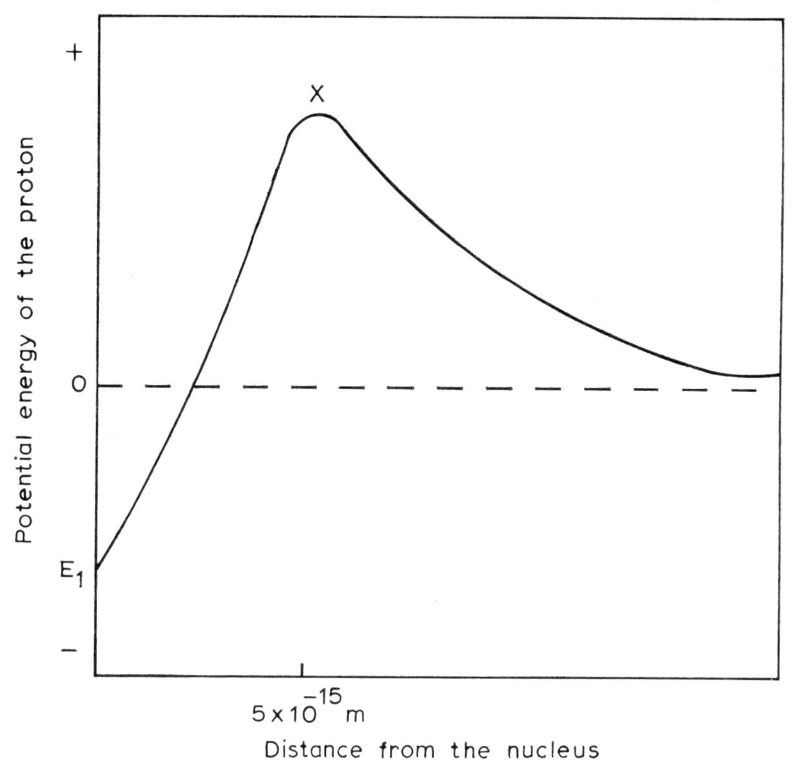

Figure 2.5. Variation in potential energy of a proton as it approaches a nucleus. A proton approaches a nucleus of small charge from an infinite distance, losing kinetic energy and gaining potential energy. If it has sufficient kinetic energy it will, according to classical mechanics, surmount the potential barrier at X and fall into the 'potential well' of the nucleus, under the action of the short range nuclear force. In this process it gains kinetic energy and loses potential energy. The kinetic energy E_1 is known as the 'binding energy'. (Reproduced from *Nuclear Fusion* by H. R. Hulme, by permission of the author.)

E_0 is large enough the particle will reach the point X when the attractive nuclear force becomes greater than the repulsive Coulomb force, and it will fall into the nucleus. In this process it gains kinetic energy and loses potential energy. Its resultant kinetic energy is now $E_0 + E_1$, and this is shared with the other nucleons with the result that the nucleus becomes excited. This excess energy must be emitted, either in the form of γ-rays, or by the expulsion of an energetic particle. When a neutron enters a nucleus it does not have to overcome any potential barrier and so low energy neutrons may be readily absorbed by certain nuclei, i.e. these nuclei have a high capture cross-section for low energy neutrons.

2.12.1. *Binding energy*. When a nucleus is formed from its respective nucleons the mass of the resulting nucleus is less than the sum of the masses of the nucleons. Consider the formation of ^4_2He. The mass of this nucleus, in atomic mass units (a.m.u.), is 4·002604 (1 a.m.u. is one twelfth of the mass of the ^{12}C nucleus, and is equal to $1{\cdot}660 \times 10^{-27}$ kg). The mass of a proton is 1·007277 a.m.u. and that of a neutron is 1·008665 a.m.u. The sum of the masses of two protons and two neutrons is:

$$2 \times 1{\cdot}007277 + 2 \times 1{\cdot}008665 = 4{\cdot}031884 \text{ a.m.u.}$$

so the loss in mass on the formation of ^4_2He from its nucleons is:

$$0{\cdot}029280 \text{ a.m.u.} \equiv 4{\cdot}8604 \times 10^{-29} \text{ kg}.$$

According to the Einstein equation, $E = mc^2$ (p. 13), this mass must be converted into energy, i.e. the energy released is:

$$4{\cdot}8604 \times 10^{-29} \times 2{\cdot}998^2 \times 10^{16} = 4{\cdot}3685 \times 10^{-12} \text{ J } (27{\cdot}3 \text{ MeV}).$$

The energy per mole is:

$$4{\cdot}3685 \times 10^{-12} \times 6{\cdot}02 \times 10^{23} = 2{\cdot}63 \times 10^{12} \text{ J}.$$

This energy released is called the *binding energy*.

Thus the binding energy of the ^4_2He nucleus is 27·3 MeV, or 6·8 MeV per nucleon. This shows the exceptional stability of the α-particle (figure 2.4).

2.12.2. *Fusion reactions*. To release energy on a large scale by the direct synthesis of ^4_2He from its nucleons is impracticable, because of the instability of the neutron, so the obvious starting material to choose is ^2_1H, commonly known as deuterium (D). As the Coulomb force of repulsion depends on the product of the charges of the two particles, it is essential to keep the charge to the minimum; hence the choice of deuterium.

The energy required to bring two deuterium nuclei together from an infinite distance is, according to Newtonian mechanics:

$$-\int_{\infty}^{r} \frac{e^2}{4\pi\epsilon_0 r^2} = \frac{e^2}{4\pi\epsilon_0 r}$$

and this is equivalent to 0·3 MeV, taking r as 5×10^{-15} m which is approximately the nuclear radius of deuterium. At this distance the nuclear force becomes greater than the Coulomb repulsion and the two nuclei will fuse. Since 1 MeV is approximately equivalent to a temperature of 10^{10} K this means that the deuterium must be heated to about 10^9 K, according to classical mechanics.

However, quantum mechanics shows that there is a finite probability of deuterons with energies considerably lower than 0·3 MeV penetrating the Coulomb barrier. This is often referred to as the 'tunnel effect'. Fusion reactions may, therefore, occur at temperatures around 10^8 K. For an explosive reaction the best method of producing such a high temperature is to use a fission bomb as a detonator and this in fact was used in the first thermonuclear explosion produced by the Americans in 1952.

The actual detailed construction of this bomb is a closely guarded military secret, but it is known that a nuclear detonator (^{235}U or ^{239}Pu) is surrounded with a compound of deuterium, such as LiD, and the whole is further surrounded with a tamper of uranium. The detonator raises the temperature to 10^8 K and initiates the following reactions:

$$^2_1D + ^2_1D \rightarrow ^3_1T + ^1_1H + 4{\cdot}03\ \text{MeV} \quad (2.4)$$
$$^2_1D + ^2_1D \rightarrow ^3_2He + ^1_0n + 3{\cdot}27\ \text{MeV} \quad (2.5)$$
$$^2_1D + ^3_1T \rightarrow ^4_2He + ^1_0n + 17{\cdot}6\ \text{MeV} \quad (2.6)$$

Reaction (2.6) goes about 100 times as fast as the other two at temperatures around 10^8 K, so the T (tritium) formed in (2.4) is very rapidly used up in reaction (2.6). Combining equations (2.4), (2.5) and (2.6) gives:

$$5\,^2_1D \rightarrow ^3_2He + ^4_2He + ^1_1H + 2\,^1_0n + 24{\cdot}9\ \text{MeV}\ (3{\cdot}98\ \text{pJ})$$

1 kg of deuterium would release:

$$3{\cdot}98 \times 10^{-12} \times 6 \times 10^{23} \times 10^{2}\dagger = 2{\cdot}39 \times 10^{14}\ \text{J},$$

which compares with 4×10^6 J released by the explosion of 1 kg of TNT. In addition the ^7_3Li would react with the high energy protons (p. 13) forming 2 ^4_2He releasing still more energy.

† 5 $^2_1D \equiv 10$ g hence the factor 10^2.

In practice only a small part of the deuterium will fuse before the parts of the bomb are separated by the explosion, but it is clear that a fusion bomb is more powerful than a fission bomb of the same weight by several orders of magnitude (see question 2.2), and 'H' bombs have been exploded equivalent in power to many megatons of TNT. There is no lower theoretical limit to the size of an 'H' bomb (except for the detonator), because, unlike a fission bomb, there is no chain reaction involved. The bomb is relatively 'clean' because fission products are only released by the detonator.

2.13. *The neutron bomb.*

This is essentially a small fusion bomb. Fusion produces about twenty times as many neutrons per g. as fission. The bomb is so designed that the energy release is much slower than a normal fusion or fission bomb and the absence of a strong blast explosion reduces very considerably the damage to buildings; fall out is limited to a small area. The high energy neutrons, about 15 MeV, released in fusion and the γ-rays will produce lethal injuries up to a range of 1–2 km. Very little, if any protection can be given against 15 MeV neutrons.

The construction of a neutron bomb is a closely guarded military secret, but it is known that a fusion bomb must be detonated by a fission bomb and the two sub-critical masses required must be driven together by a chemical explosive. Fissionable nuclei heavier than ^{239}Pu may be used, which have a higher neutron output per fission. This could reduce the size of the bomb considerably.

Questions

2.1. Calculate the energy which could be liberated from 1 mole of uranium on the fission of half of the ^{235}U into ^{137}Xe, ^{95}Ru and 3 neutrons. Compare this to the heat of combustion of carbon as graphite (395 kJ per mole). (Percentage abundance of $^{235}U = 0{\cdot}07$, nuclear masses of $^{235}U = 235{\cdot}067$, $^{137}Xe = 136{\cdot}956$, $^{95}Ru = 94{\cdot}943$, $^{1}n = 1{\cdot}009$, $c = 3 \times 10^{8}\,m\,s^{-1}$.)

2.2. Calculate the energy released in the fission of 1 kg of ^{235}U and compare this with the energy released in the explosion of 1 kg of TNT (4×10^{6} J). (Energy released per fission of 1 atom of $^{235}U = 200$ MeV $= 32$ pJ.)

2.3. Calculate the time required for the fission of all the atoms in 50 kg of ^{235}U (the approximate mass of a fission bomb). (Average energy of neutrons released in fission $= 1{\cdot}5$ MeV, average number of neutrons released per fission = 2·5, mass of the neutron = $1{\cdot}67 \times 10^{-27}$ kg.) (Notes on the calculation: Determine the total number of neutrons released and their velocity and hence the time between collisions. Use equation 2.2 (p. 26), taking $N_0 = 1$ and $k = 2$. Assume that the atoms of ^{235}U do not separate during the time of the explosion and that every collision between a neutron and a ^{235}U nucleus is effective.)

3. The development of nuclear reactors

3.1. *Introduction*

The origin of nuclear power appears mysterious; however, there are similarities between nuclear and conventional fuels. Both involve the mining of raw materials, their processing and transport to power stations for combustion to produce heat, which raises steam to drive turbines for generating electricity. Nuclear fission is not strictly a combustion process, though the term 'burn up' is commonly used; the uranium reacts with neutrons rather than oxygen. Both types of fuel produce waste products which can be harmful.

Though the first reactor was not made until 1941, it now seems reasonably certain that a *natural* chain reaction started spontaneously at Oklo in the Gabon, West Africa, many millions of years ago. At that time the ratio of ^{235}U to ^{238}U was much higher than it is today, and the chain reaction was probably started by water, acting as a moderator, percolating into the uranium ore. The discovery was made accidentally a few years ago, when a French company was mining for uranium in the Gabon. When the uranium was charged into a reactor the chain reaction would not start. Analysis showed that the ^{235}U content was much below the normal 0·7%. From the observed ratio of ^{235}U to ^{238}U it was concluded that this natural reactor had operated at an average of 20 kW for some 10^5 years.

Apart from the military applications, the main effort since 1945 has been directed to producing improved types of nuclear reactor. Reactors are used in research, and for supplying electrical energy and producing radioactive isotopes; they also produce plutonium as a valuable by-product. The first British reactor was built at Harwell in 1947 and called GLEEP (Graphite Low Energy Experimental Pile). It is still operating and has a maximum flux of 10^{10} neutrons $cm^{-2}\ s^{-1}$. GLEEP was soon followed by the larger reactor BEPO, now closed. Both of these were experimental reactors to test materials for the atomic energy programme, though BEPO was used for the production of radioactive isotopes. Other experimental reactors were built, such as LIDO and HERALD, which use ordinary water as a moderator and uranium highly enriched in ^{235}U as the fuel. These reactors are comparatively cheap to construct. Still others, such as DIDO and PLUTO at Harwell, use heavy water (deuterium oxide) as moderator, and uranium, highly enriched in ^{235}U or ^{239}Pu, as the fuel. Natural uranium can be used with heavy water as moderator owing to the low capture cross-section of deuterium compared with that of hydrogen. The heavy water reactors at Harwell are now used in place of BEPO for the production of radioactive isotopes.

3.2. *Dimensions and construction of a reactor*

A typical air-cooled graphite moderated uranium reactor would have a core consisting of a stack of graphite blocks about 18 m across pierced by a large number of channels into which the fuel elements can be inserted, at intervals of 15 to 20 cm (Figure 3.1). The core of a heavy water moderated reactor would be only 3–5 m across (Figure 3.3). The light water reactor, which uses uranium highly enriched in ^{235}U, has an even smaller core. The dimensions of the core and the spacing of the fuel elements are determined by the average thickness of the moderator required to reduce a fission neutron to thermal energy—about 6 cm for light water, 11 cm for heavy water, and 18·5 cm for graphite. The neutrons follow a very erratic path. In graphite the average distance between collisions is 2·5 cm, and as the neutron loses one-sixth of its energy per collision it makes approximately 100 collisions before reaching thermal energy, travelling a distance of about 250 cm.

As mentioned on p. 27 an average of 2·5 neutrons per atom are released in the thermal fission of ^{235}U. In pure natural uranium, owing to resonance capture, the number (η) available to produce further fissions of ^{235}U is <1, and so a chain reaction is impossible. In a perfectly moderated natural uranium reactor η is 1·3, and this is the maximum possible value of the reproduction constant k_∞ for a reactor of infinite size (see Equation 2.1). In practice k_∞ rarely exceeds 1·07 owing to the reaction of thermal neutrons with the moderator and with impurities. It is essential to design the reactor so that k decreases with increase in temperature (p. 46). Most reactors are designed, as explained on p. 25, with a k factor of about 1·02, so that as a result of the build-up of fission products and burn-up, i.e. fission of fuel, the control rods can be drawn out to keep the value of k steady at 1·000. Some of the secondary fission products have high values of σ_a (the capture cross-section), notably ^{135}Xe, for which $\sigma_a = 2{\cdot}7 \times 10^6$ b. This particular isotope

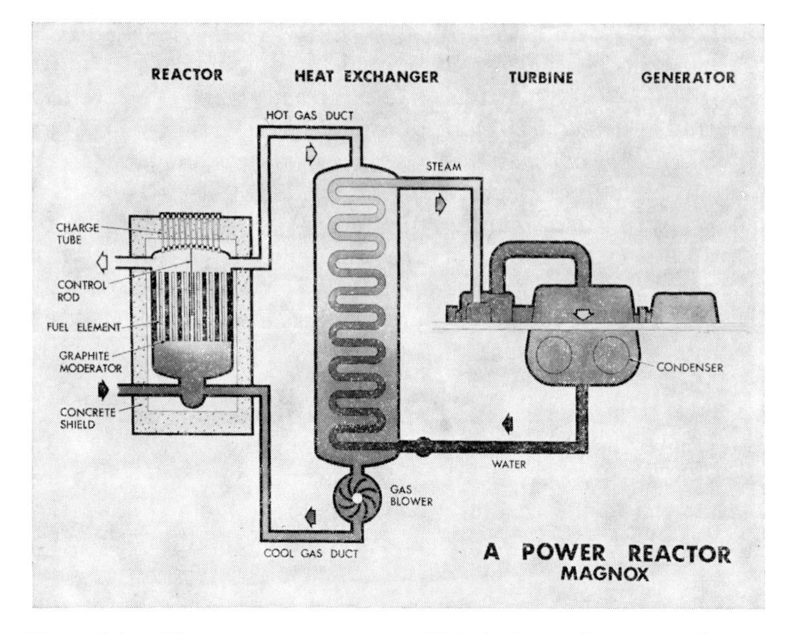

Figure 3.1. Magnox power reactor. This is the earliest type of reactor used for producing electric power. Calder Hall was the first one to be built in Britain. It is graphite moderated, the natural uranium metal is clad in a magnesium alloy (Magnox) which limits the working temperature to 400°C. CO_2 is used as coolant. The core is about 18 m across and is surrounded by graphite as a neutron reflector. An improved version of this reactor is the Advanced Gas Cooled Reactor (AGR). This uses uranium(IV) oxide, enriched in ^{235}U to 3%, clad in stainless steel. It is more efficient than the early Magnox stations and operates at 500°C. (Figure supplied by UKAEA, who retain copyright.)

has a half-life of 9·2 h, and as it is a secondary product formed by the decay of ^{135}I (half-life 6·7 h) the concentration in the reactor increases for several hours after a shut down. This factor may prevent the reactor starting up again until the ^{135}Xe has decayed to a sufficiently low level. It has been suggested that the problem might be overcome by surrounding the core with a layer of beryllium. Neutrons can be released from beryllium by a γ, n reaction with γ-rays in excess of 1·7 MeV energy. Owing to the high level of γ-radiation from the fission products enough extra neutrons would be produced by the γ, n reaction in the beryllium to enable the reactor to be restarted, despite the ^{135}Xe ‘poisoning’.

The core of a reactor is surrounded by a ‘reflector’ about 1 m thick, which again consists of graphite blocks. The function of the reflector is to reduce the escape of neutrons and to turn back a considerable proportion into the core. Finally, the reflector is surrounded by a biological shield to absorb excess neutrons, and also to absorb the very considerable γ-radiation given out during fission and during the decay of the primary fission products. The shield is made of concrete and is about 3 m thick. At one or more points the graphite is allowed to penetrate the biological shield to form a 'thermal column'. This is a region remote from the core where thermal neutrons exist mixed with a very low proportion of fast neutrons, and is used for experimental purposes. The face of the thermal column is covered with cadmium sheet to absorb thermal neutrons and a thick layer of lead to reduce the γ-radiation to a safe level.

3.3. *Power reactors*

3.3.1. *Thermal reactors.* The first British nuclear power station was opened in 1956 at Calder Hall to run at 180 MW, subsequently increased to 250 MW. This reactor is graphite moderated and CO_2 is used as coolant to heat water in the heat exchanger (Figure 3.1). The uranium metal is clad in a magnesium alloy called Magnox, and this limits the operating temperature to 400 °C. A number of similar reactors (Magnox) were built, the design being improved each time. They suffer from the common disadvantage of all natural uranium thermal reactors, that only a very low percentage of the 'fuel' can be used, since the fissile isotope ^{235}U is only present at 0·7 %.

An improved version of this reactor is the Advanced Gas Cooled Reactor, or AGR. This uses uranium(IV) oxide clad in stainless steel, which enables the operating temperature to be about 500 °C, giving increased efficiency. Owing to the higher value of σ_a for iron as compared with magnesium the uranium must be slightly enriched in ^{235}U, from 0·7 up to 3 %.

A third generation of gas cooled reactors (HTR, Figure 3.2) uses helium as coolant, and the uranium(IV) oxide is encased in carbon or silicon carbide in the form of small pellets, unlike the fuel elements used in the Magnox reactors and the AGR, which are in the form of finned rods. The fission products are contained within the coating. Still higher temperatures are possible with this reactor.

3.3.2. *Wigner energy.* One of the problems associated with graphite moderated reactors is the deformation of the graphite lattice by continuous neutron bombardment. This results in stored energy called 'Wigner energy', after E. P. Wigner, who predicted it. This Wigner energy must be released periodically by allowing a rise of temperature to take place. The rise of temperature has to be carefully controlled as a further rise of temperature will occur when the carbon atoms return to their natural position. An accident occurred at Windscale during a Wigner release. The temperature rose very rapidly and ten tonnes of uranium became incandescent, scattering fission products from the smoke stack over the local countryside. Fortunately, the level of contamination did not prove serious, as filters had been fixed in the smoke stack.

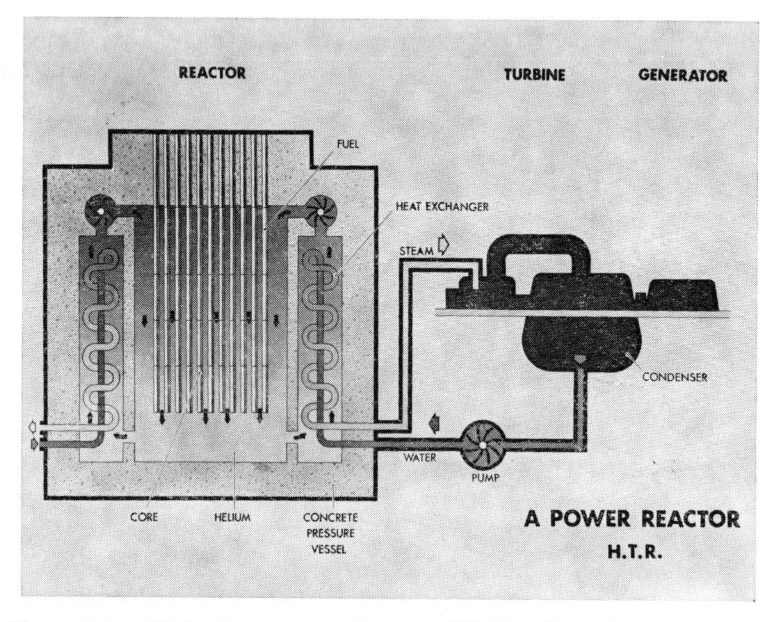

Figure 3.2. High Temperature Reactor (HTR). Few, if any, of these reactors have been so far used for power production. The fuel is enriched uranium(IV) oxide encased in carbon or silicon carbide in the form of pellets and helium is used as coolant. No additional moderator is required. The reactor is more efficient than the Magnox or AGR reactor as still higher operating temperatures are possible. (Figure supplied by the UKAEA, who retain the copyright.)

Other types of thermal reactors use either light water or heavy water as moderator, and light water or heavy water as coolant. There are two types of light water reactor (LWR), the pressurised water reactor (PWR) and the boiling water reactor (BWR). The main development in the USA has been the PWR and this is used in nuclear powered submarines. The PWR uses a heat exchanger system, but the BWR produces steam direct to drive turbines. Both LWRs have a lower thermal efficiency than the AGR.

The heavy water system has been mainly developed in Canada and has been given the code name of 'CANDU'. CANDU uses heavy water, both as moderator and coolant, and uses a heat exchanger. A different type has been developed in Britain, using light water as coolant which is converted to steam. This is known as the SGHWR (Figure 3.3).

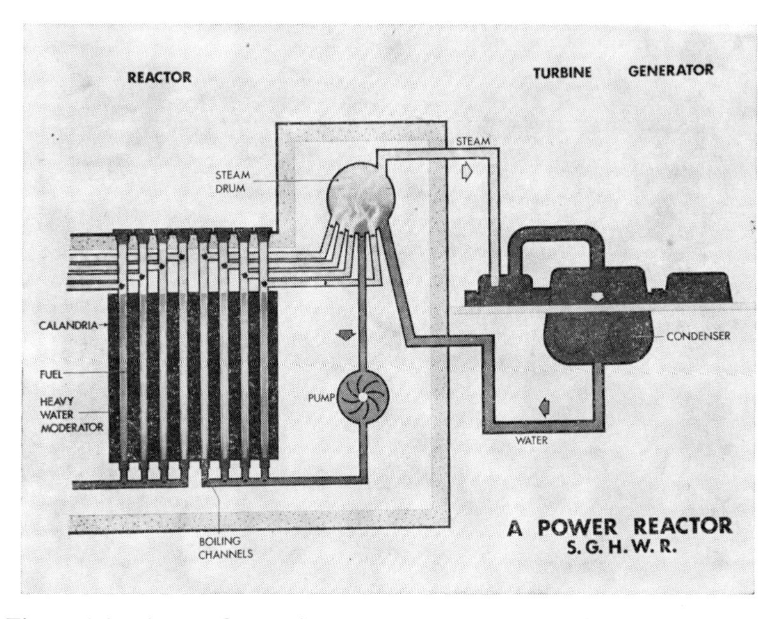

Figure 3.3. Steam Generating Heavy Water Reactor (SGHWR). This reactor uses natural, or slightly enriched, uranium as fuel, and heavy water as moderator enclosed in a sealed vessel. Heavy water is an ideal moderator and so the core is much smaller than either Magnox or AGR. Light water passes over the fuel rods as coolant and is converted to steam which drives the turbines. There is no heat exchanger. (Figure supplied by the UKAEA, who retain copyright.)

The size of the heavy water system is somewhat larger than the LWR, but much smaller than the Magnox or the AGR. Deuterium is an ideal moderator with a capture cross-section of 0·00057. It is interesting to reflect that Hitler adopted heavy water as moderator for the reactor designed in Germany during the war, but this reactor never became operative.

3.3.3. *High Flux Reactor.* This reactor, designed and built in the USA, is intended to produce trans-plutonium elements by multiple neutron capture. It consists of a cylinder of water surrounded by highly enriched uranium. This in turn is surrounded with a neutron reflector of beryllium. The central cylinder of water acts as a ' neutron trap ' because the lifetime of neutrons in the water is appreciably longer than in the uranium. Fluxes of the order of 5×10^{15} n cm^{-2} s^{-1} can be obtained in the cylinder. Plutonium, americium or curium obtained by irradiation in a normal reactor are placed in the cylinder and irradiated for long periods. The reactor is housed in a pressure vessel and rests at the bottom of a deep tank of water (see p. 55).

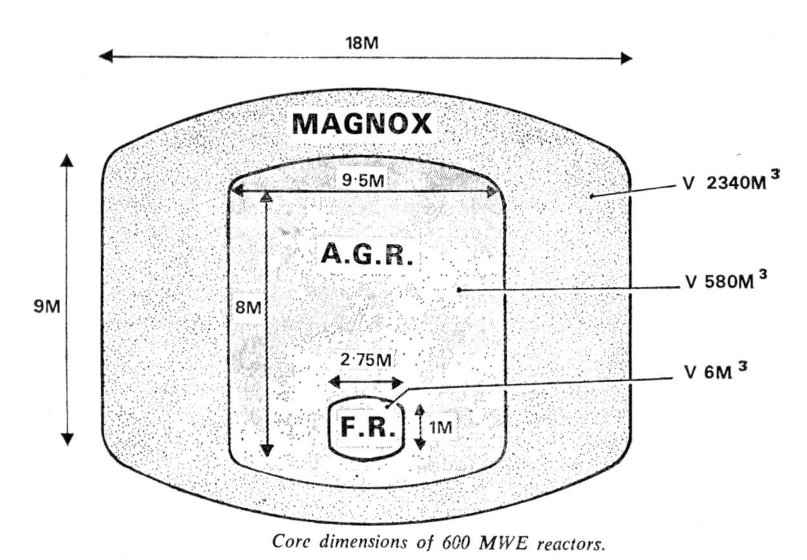

Figure 3.4. Core dimensions for 600 MWE reactors. (Figure supplied by the UKAEA, who retain copyright.)

3.3.4. *Fast Breeder Reactor* (*FR*). This type of reactor, which has been developed in Britain, France, U.S.S.R. and America, offers great possibilities for the future. The British design at Dounreay, which went critical in 1958, has now been replaced by the FR. This is probably the most advanced reactor in the world. It operates by the fast neutron reaction; no moderator is used, and the volume of the core is very small (Figure 3.4). Plutonium is the most satisfactory fuel as it produces an average of 2·9 neutrons per fission compared with 2·4–2·5 for ^{235}U. Owing to the large energy release in a small volume, very efficient cooling is required, and liquid sodium is used as coolant. For efficient heat transfer thin tubes are necessary. Traces of oxygen in the metal pipes produce sodium oxides, which greatly increase the rate of corrosion and also tend to block the pipes so that separators have to be incorporated in the circuit. A secondary liquid sodium circuit is used for heat transfer, and this in turn converts water to steam to drive the turbines. The reactor operates at about 600°C (Figure 3.5).

The core of the reactor is very compact. Figure 3.4 shows the size of the core compared to thermal reactors. The core weighs about 5 tonnes, consisting of 20% plutonium intimately mixed wth uranium. 70% burn-up is possible compared with about 0·5% for a Magnox reactor. The core produces one hundred times as much heat as the same volume of a thermal reactor.

The supreme advantage of this reactor is that it can 'breed' its own fuel. The core is surrounded by a blanket of natural uranium which captures spare neutrons to produce more plutonium. Natural thorium can also be used and is converted to fissile ^{233}U.†

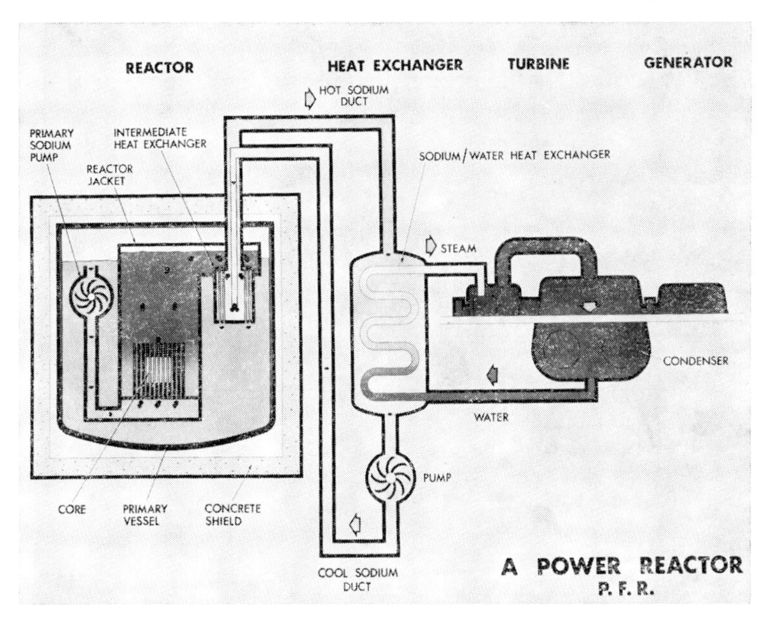

Figure 3.5. Fast Breeder Reactor (FR). This operates by the fast neutron reaction, no moderator being used. The fuel is ^{239}Pu intimately mixed with 80% uranium. The core is very small (Figure 3.4) and is surrounded by natural uranium, so that the excess neutrons released in fission produce more plutonium fuel. The coolant is liquid sodium which becomes intensely radioactive, so that an intermediate heat exchanger with a secondary sodium loop is used, within the primary vessel. This secondary sodium loop raises steam. This reactor is by far the most efficient so far produced and must be the preferred reactor of the near future. (Figure supplied by the UKAEA, who retain copyright.)

† Opposition to the construction of the FR is particularly misguided because the FR is a plutonium incinerator.

3.3.5. *Safety of reactors.* Reactors in general are slowed down automatically with a rise of temperature due to a fall in density of the moderator, though changes in cross-section with temperature may increase the rate, notably in the case of plutonium. Allowances must be made for this in reactor design.

If the moderator is lost, as could happen in the case of light or heavy water reactors, the reactor will stop. However, when the chain reaction stops, heat is still being generated from the decay of fission products. The main hazard is the failure of the cooling system. The possible loss of coolant is a much more serious hazard with the compact LWR. In these reactors loss of moderator means loss of coolant. If the cooling system failed the fission product heating could lead to a rapid rise of temperature and melting of the core could occur, releasing large quantities of fission products. It is therefore essential to provide an emergency cooling system. The PWR presents the greatest hazard in this respect, because of its small size, and because it is under high pressure. The FR with its extremely compact core presents a considerable hazard from failure of the cooling system, but it is not under pressure.

3.3.6. *Efficiency of reactors.* The following table gives a comparison of the efficiency of some types of reactor. The data is based on a reactor operating at 1000 MW.

	Magnox	AGR	BWR	FR	
Peak fuel rating (MWH/tonne of fuel)	3	14	27	260	
Fuel feed U_3O_8 (tonnes/year)	370	200	240	1	
Fuel utilization (%)	0·2	0·4	0·4	70	
Plutonium yield (tonnes/year)	0·63	0·22	0·29	0·26	nett

It will be clear from this table that the fast reactor has very great advantages compared to other types. It has been estimated that, on a world scale, thermal reactors could be consuming 50 000–100 000 tonnes of uranium per year by the mid 1980s. The estimated world reserves of natural uranium high grade ore are between 200 000 and 500 000 tonnes, so that these would be exhausted before the end of the century and low grade ores would have to be used. They are much more costly to work. The fast reactor overcomes this problem owing to its greater efficiency, and should reduce the cost of power production. The engineering and development costs, however, are high.

3.3.7. *Initial charge of plutonium for the FR.* The chief source is the spent fuel rods of the Magnox reactors. Most of the plutonium is ^{239}Pu, but considerable quantities of ^{240}Pu and ^{242}Pu are also produced. Both of these isotopes are fissionable by fast neutrons and can be used in the FR only. Some burn-up of ^{239}Pu occurs in Magnox and AGR reactors.

3.3.8. *Chemical processing.* Owing to the burn-up of fissionable material and to the build-up of fission products, many of which have high capture cross-sections, the fuel elements have to be withdrawn periodically and replaced. They are highly active and have to be stored for a period of time in adequate shielding. They are then taken to a chemical separation plant, where the cladding must first be removed. This may be done mechanically, or the entire fuel element may be dissolved in nitric acid to produce an aqueous solution of uranium, plutonium and fission products. The lanthanum fluoride and bismuth phosphate cycles used in the war-time extraction plant have now been superseded by solvent extraction methods. Tributyl phosphate (TBP) is generally used to separate the fission products from uranium and plutonium. It has several advantages. Firstly, solvent extraction is easier to operate by remote control than are precipitation processes. Secondly, TBP is relatively stable to radiation. Thirdly, the partition coefficient for U(VI), Pu(VI) and Pu(IV) between TBP and nitric acid of suitable concentration is very high, the fission products remaining in the aqueous phase. These fission products are a valuable source of radioactive isotopes, and some of them such as ^{131}I, ^{137}Ba, ^{137}Cs and ^{90}Sr are separated for future use.

To separate plutonium from uranium use is made of the low partition coefficient for Pu(III) between TBP and aqueous solutions. Plutonium can be reduced to Pu(III) with sulphur dioxide or iron(II) sulphamate, which leave the uranium as U(VI) in the organic phase, while the plutonium is extracted into aqueous solution.

In all these operations great care has to be taken to prevent the concentration of fissionable material becoming critical. For fast breeder reactors there is no need to separate the plutonium completely from uranium. For instance, the solution might be adjusted to 20% plutonium, then both uranium and plutonium could be precipitated as hydroxides with ammonia. This process ensures the dispersion of the plutonium in natural uranium so that the mass cannot become critical.

3.3.9. *Disposal of radioactive waste.* This constitutes a major problem. Most of the fission products have no uses at present. They must either be stored in concentrated solution in tanks below ground level, or converted into ceramic materials, where they remain in permanent storage. Deep, disused mines may also be used in future. The bulk of the radioactivity will eventually decay, leaving nearly inactive clays. In the processing plants, slightly active wastes are generated in large volumes of water, and these wastes are disposed of at sea under very carefully controlled conditions. The problem of waste disposal, particularly in connection with the proposed reprocessing plant at Windscale in the U.K., has aroused considerable controversy.

3.4. *The possibility of a fusion reactor*

The explosive release of energy from fusion was explained in the last chapter. It is now generally accepted that the energy given out by the sun and stars is produced by continuous fusion reactions. Is it possible to reproduce these reactions on Earth by the controlled release of fusion energy? The idea is attractive, because the most suitable raw material, deuterium, is available in almost unlimited quantity (0·017% of water is D_2O), unlike uranium. The most suitable reaction to use is (p. 35):

$$^{2}_{1}D + ^{3}_{1}T \rightarrow ^{4}_{2}He + ^{1}_{0}n + 17{\cdot}6\ MeV \qquad (3.1)$$

In principle, a plasma (i.e. positive nuclei mixed with electrons) must first be produced, and then the plasma must be raised to 10^8 K so that the above reaction occurs.

The energy given out will be in three forms:

(*a*) kinetic energy of the charged particles,
(*b*) kinetic energy of the neutrons,
(*c*) electromagnetic radiation.

(*a*) must be retained in the plasma to maintain the fusion temperature and to heat up new fuel as it is injected. (*b*) and (*c*) must be absorbed in the containing walls and extracted with heat exchangers as useful energy. Three-quarters of the energy given out will be in the form of (*b*) and (*c*). The walls must be made of deuterium-rich material so that the neutrons will 'breed' tritium, which is the really expensive fuel. The breeding ratio, i.e. the ratio of tritium generated to tritium consumed, should be greater than 1, so that more tritium is formed to start new reactors (Figure 3.6).

The chief problems to be overcome are the instability of a plasma which causes it to fly to the walls of the container and lose its heat, and the injection of new fuel. In the idealised diagram of a fusion reactor (Figure 3.6) the plasma is shown surrounded by a vacuum, i.e. the plasma must be contained in the centre. To achieve this a very strong magnetic field must be applied which causes the particles to rotate about the flux lines in helices. Collisions produce movement across the flux lines, with resultant diffusion out of the field.

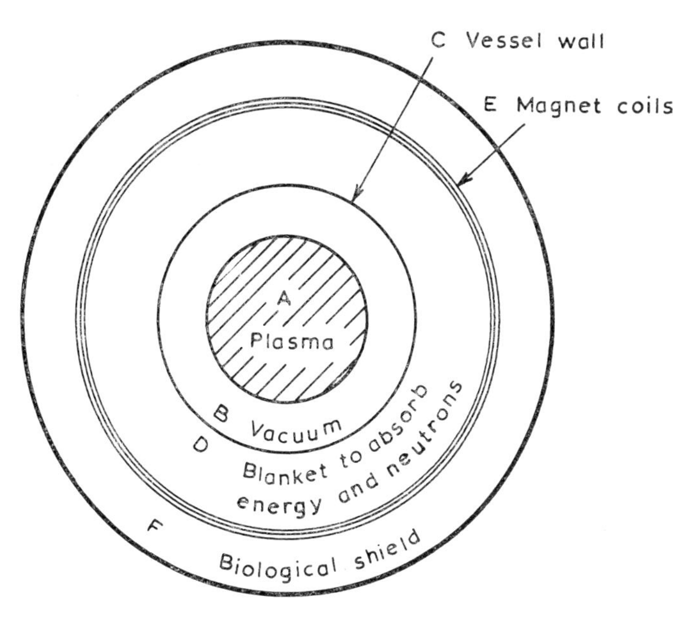

Figure 3.6. The principle of a fusion reactor. The plasma A must be contained in the centre of the vacuum vessel B by means of a magnetic field. The vessel B is surrounded with a material rich in deuterium. This performs two functions: (i) as a heat exchanger, by absorbing the energy of the neutrons and using this energy to raise steam for turbines, (ii) to ' breed ' tritium for fusion fuel by the reaction:

$$^{2}_{1}D + ^{1}_{0}n \rightarrow ^{3}_{1}T + \gamma + 6{\cdot}3 \text{ MeV}$$

(Reproduced from *Nuclear Fusion* by H. R. Hulme, by permission of the author.)

The Thetatron. In this apparatus a plasma is first produced by a relatively small longitudinal current and then a very large current is passed by the discharge of a large bank of capacitors, and a very powerful magnetic field is generated inside the plasma. The magnetic field generates a so-called θ current in the opposite direction to the original current, and this counter-current must be sufficient to produce a magnetic field which almost neutralizes the applied field in the *centre* of the plasma. This leaves the residual magnetic field *outside* the plasma which exerts a considerable pressure and drives the plasma into the centre of the tube. This is called the 'pinch' effect. The particles acquire very high energies due to the pressure, and temperatures of the order of 10^7 K have been obtained by this means, but the plasma diffuses out at the two ends of the magnetic field. This diffusion can be reduced by arranging two coils at either ends of the tube to form a type of magnetic mirror. There is some evidence that thermonuclear reactions have been momentarily produced in certain experiments with the thetatron, but it does not seem to be a very promising method for the development of a continuous process because it only lasts for a few microseconds. However, the apparatus gives much valuable information about the behaviour and properties of a plasma.

In order to avoid the diffusion at the ends of the magnetic field in a linear apparatus, some workers in America and Russia have developed a circular apparatus, or torus (Figure 3.7). This is known as a 'Stellarator' by the Americans, and a 'Tokamak' by the Russians. The torus has its own problems, but, in general, it may be said that this offers the most hopeful line of advance at the present time.† How to keep the plasma stable for a sufficient time at a temperature high enough to produce a continuous thermonuclear reaction is the outstanding problem.

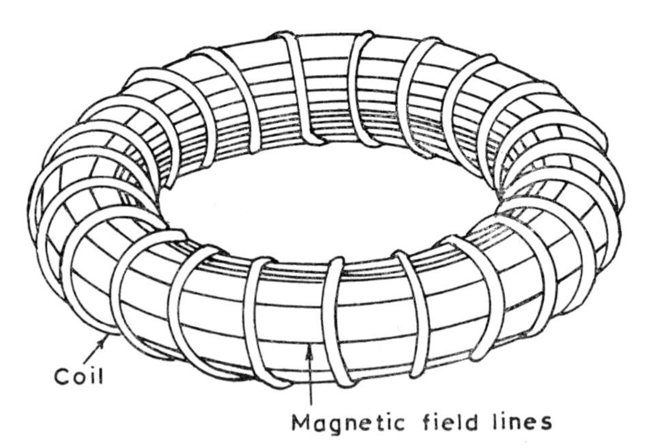

Figure 3.7. A toroidal magnetic trap. A closed system, in the form of a torus, seems at present to offer the best prospect of realizing a continuous thermonuclear reaction. A powerful magnetic field contains the plasma in the centre and a heat exchanger and biological shield surround the torus. The very difficult problem of injecting fresh fuel to sustain the reactor has yet to be solved. (Reproduced from *Nuclear Fusion* by H. R. Hulme, by permission of the author.)

† JET (Joint European Torus) is now under construction at Culham.

4. The actinides

4.1. *Introduction*

The discovery of neptunium and plutonium as outlined in Chapter 2, p. 1a, raised some interesting possibilities. Were these elements part of a 6d transition series, or were they part of a new rare earth or 5f series, and if so where did the 5f series start? It had been assumed before 1940 that the elements actinium, thorium, protoactinium and uranium were the first four members of a 6d series. In their oxidation states they showed considerable resemblance to lanthanum, hafnium, tantalum and tungsten, respectively. If neptunium and plutonium were members of a 6d series they should resemble rhenium and osmium. Rhenium has seven oxidation states: +1, +2, +3, +4, +5, +6 and +7, resembling manganese closely. Osmium has five: +2, +3, +4, +6 and +8 and shows a marked resemblance to ruthenium. Neptunium and plutonium have four principal oxidation states: +3, +4, +5 and +6, showing a close resemblance to uranium and little resemblance to rhenium and osmium. If a comparison is made between neptunium and plutonium and the corresponding lanthanides, promethium and samarium, which show the typical +3 oxidation state of the lanthanides, the resemblance is far from obvious. The chemical evidence so far is inconclusive. However, when the magnetic and spectroscopic data are considered there is strong evidence for the presence of f electrons in the ground states of protoactinium, uranium, neptunium and plutonium. The synthesis of elements beyond plutonium confirmed the resemblance to the lanthanides and the group is now known as the actinides.

The study of these elements demanded the development of new chemical techniques, either the use of conventional size apparatus and concentrations as low as 10^{-12} M, detecting the elements by their radioactivity, or the use of ultramicrochemical techniques with microgram quantities and concentrations of 10^{-1} to 10^{-3} M; this involved volumes of 10^{-3} cm^3 to 10^{-5} cm^3 and observation of results under a microscope. Further problems were encountered owing

TABLE 4.1. *Actinides* 89–96.

Element	Discovery and occurrence	Preparation, isolation or extraction	Isotopes	Oxidation states	Points of chemical interest	Uses
$_{89}$Ac	Debierne, 1899. In lanthanide fraction of pitchblende. [3] Occurs as 1 part in 10^{10}	By solvent extraction or ion exchange from pitchblende.	Natural—^{227}Ac, ^{228}Ac. 13 artificial. Half-lives vary from $3{\cdot}9\times10^{-4}$s to 22y.	+3	Compounds more basic than those of the corresponding lanthanide, La, because of the large size of Ac^{3+}.	—
$_{90}$Th	Berzelius, 1828. Occurs principally in monazite sand, but widely distributed in Earth's crust.	Extracted by cation exchange. Th^{+4} held strongly. Eluted as an oxalate complex. (See Expt. 27.)	6 natural, 12 artificial. Half-lives vary from 5×10^{-4}s to 10^{10}y.	+4. +2, +3 also exist.	Typical Group IV element.	Original: for incandescent gas mantles. Modern: for preparation of fissile ^{233}U.
$_{91}$Pa	Fajans, 1913, as ^{234}Pa in natural uranium. ^{231}Pa ($t_{1/2}=10^4$y) occurs in uranium minerals as the third member of the $(4n+3)$ series.	^{231}Pa—by neutron irradiation of ^{230}Th. Extracted from conc. HCl by organic solvents. (See Expt. 7.)	2 natural, 13 artificial. Half-lives vary from 0·6s to 10^4y.	+5 +4 also exists.	Ions very easily hydrolysed, forming colloid particles which are carried down by any precipitate. Chemistry very difficult to study.	—
$_{92}$U	Klaproth, 1789. In pitchblende. Metal isolated in 1841. Occurs at 4 ppm in Earth's crust. Principal ores are pitchblende, carnotite, autunite. Oceans estimated to contain 10^{10} tonnes. Named in honour of Herschel's discovery of Uranus.	Extracted by flotation followed by roasting and acidic or alkaline leaching. Purified by anion exchange as complex chloride, or by solvent extraction using tributyl phosphate (TBP); partition coefficient in high NO_3^- concentration is 2000 : 1.	Natural—^{238}U, ^{235}U, ^{234}U. 11 artificial. Half-lives vary from 1·3 min to 10^9y.	+3, +4, +5, +6. +6 most stable, +5 least stable. +3 very powerful reducing agent (reduces H_2O to H_2).	Strongly electropositive metal. Reacts with oxygen slowly at room temperature. Forms a wide range of complex ions in +6 state. Only volatile compound UF_6.	Original use for making coloured glass. ^{238}U generally enriched with ^{235}U in thermal reactors. ^{233}U also fissile and may be used in fast reactors of the future.
$_{93}$Np	^{239}Np—McMillan and Abelson, 1940 (see p. 19.) ^{237}Np occurs in uranium ores at 1 part in 10^{15} (compare ^{226}Ra 1 part in 10^7). Named after planet Neptune.	^{237}Np prepared in fast reactors: $^{235}U+{}^1n\rightarrow{}^{236}U+{}^1n\rightarrow$ $^{237}U\rightarrow{}^{237}Np+\beta^-$ Separated from U, Pu, and fission products by oxidation reduction cycles and solvent extraction, or ion exchange. (See Expt. 27.)	^{237}Np occurs naturally; 11 artificial. Half-lives vary from 4 min to 10^6y.	+3, +4, +5, +6, +7. +5 is the most stable state (contrast U). Lower oxidation states much more stable than those of U.	Reduced to Np(V) and Np(IV) by SO_2 or N_2H_4, and rapidly to Np(IV) by Fe^{2+}. MnO_4^- in acid oxidizes any state to Np(VI).	—

TABLE 4.1.—*Continued.*

Element	Discovery and occurrence	Preparation, isolation or extraction	Isotopes	Oxidation states	Points of chemical interest	Uses
$_{94}$Pu	^{238}Pu—Seaborg, late 1940, by deuteron bombardment of ^{238}U via ^{238}Np as intermediary [21]. ^{239}Pu—1941, by neutron bombardment of ^{238}U. ^{239}Pu occurs in uranium ores at 1 part in 10^{14}.	^{239}Pu obtained in quantity by neutron capture by ^{238}U in thermal or fast reactors. Separated and purified by oxidation/reduction cycles and solvent extraction as Pu(IV) using TBP. Separated from U by reduction to Pu(III) (insoluble in TBP).	^{239}Pu is only naturally-occurring isotope (minute amounts); 14 artificial. Half-lives vary from 20 min to 10^8y.	+3, +4, +5, +6, +7. +4 the most stable state. Pu(III) more stable than Np(III) or U(III).	+3, +4, +5, +6 states can all coexist in aqueous solution—a unique property. PuF_6 is the only volatile compound—one of the most poisonous known substances. Colours of ions remarkable. Pu^{+3} brilliant blue, Pu^{+4} bright green, Pu^{+6} yellow or orange. An element of fearful destructive potential, and of enormous potential as a source of energy.	In fast reactors and in fission and fusion bombs ^{238}Pu($t_{1/2}$=90y) is a useful heat source in electric power systems of Satellites. When mixed with Be, a useful neutron source with a low γ background.
$_{95}$Am	^{241}Am—Seaborg, 1945 [22], by multiple neutron capture from ^{239}Pu followed by β-decay of ^{241}Pu (fig. 4.1). Named after continent of origin (cf. Eu, the corresponding lanthanide).	^{243}Am (the most useful isotope)—by intense neutron bombardment of ^{239}Pu (fig. 4.1).	No natural isotope; 13 artificial. Half-lives vary from 9×10^{-4} s to 7400 y (^{243}Am).	+3, +4, +5, +6. +3 the most stable (compare U, Np, Pu).	Stability of +3 state in HNO_3 solution is basis of separation from lower actinides. Preliminary separation by anion exchange using 7M HNO_3, then eluted from cation exchange column using 5 M NH_4CNS. Np(V), U(VI), and Pu(VI) come off in first two volumes followed by Am(III). Fission products follow later.	^{241}Am mixed with Be forms a good neutron source with a very low γ background.
$_{96}$Cm	Seaborg *et al.* 1944, by α-particle bombardment of ^{239}Pu [23]: $^{239}Pu+^{4}He\rightarrow^{242}Cm+^{1}n$ Named after Marie Curie (cf. Gd, the corresponding lanthanide, named after a pioneer of lanthanide studies).	^{247}Cm ($t_{1/2}$=10^7y) prepared by prolonged neutron bombardment of lower Cm isotopes in a high flux reactor (p. 43), Separated from Pa, Am, and fission products by cation exchange, eluting with ammonium 2-hydroxy-2-methyl-propanoate.	13 known. Half-lives vary from 2·5 h to 10^7 y.	+3	Chemistry simple—only +3 state known. Stability owing to half-complete 5f shell.	^{242}Cm and ^{244}Cm used in radionuclide batteries. Both isotopes now produced in kg quantities

TABLE 4.2. *Actinides* 97–100.

Element	Discovery and occurrence	Preparation, isolation or extraction	Isotopes	Oxidation states	Points of chemical interest	Uses
$_{97}Bk$	1949, by bombarding ^{241}Am with α-particles [24]: $^{241}Am + ^{4}He \rightarrow ^{243}Bk + 2^{1}n$ ^{243}Bk has $t_{1/2} = 4{\cdot}5$ h. Named after city of discovery.	Lighter, neutron-deficient isotopes prepared by α, n (p. 72) reaction on Am or Cm. Heavier isotopes prepared by prolonged neutron bombardment of ^{239}Pu or ^{243}Am in High Flux Reactor (fig. 4.1).	9 known. Half-lives vary from 57 min to 10^{3}y (^{247}Bk).	+3, +4.	+3 is normal oxidation state; +4 obtained only with powerful oxidizing agents. Bk separated on a cation column and eluted with 2-hydroxy-2-methylpropanoate (fig. 4.3).	—
$_{98}Cf$	1950, by method similar to that for Bk, but using ^{242}Cm [25]. Named after state in which discovered.	By bombarding ^{238}U with C^{6+} accelerated to 100 MeV. $^{238}U + ^{12}C \rightarrow ^{249}Cf + ^{1}n$ Heavier, more stable, neutron-rich isotopes prepared as for Bk.	15 known. Half-lives vary from 3·4 min to 3×10^{3}y.	+3, but other, unstable, states have been reported.	Recently CfF_4 (green) has been prepared. Some evidence for Cf^{2+}. ^{252}Cf decays mainly by α-emission, but 3 per cent of decay events are spontaneous fission.	^{252}Cf a useful, highly compact, neutron source. ($2{\cdot}3 \times 10^{12}$ n s^{-1} g^{-1}). (p. 141.)
$_{99}Es$ and $_{100}Fm$	1952—both elements first detected in debris from first thermonuclear explosion [26]: collected on filter paper by drone planes flying through the mushroom cloud. Named after Einstein and Fermi.	By irradiation of ^{239}Pu in High Flux Reactor (fig. 4.1). Best yields from underground thermonuclear explosions.	*Es*: 11 known. Half-lives vary from 1·3 min to 270d. *Fm*: 14 known. Half-lives vary from $3{\cdot}3 \times 10^{-3}$s to 80d.	+3. No others so far recorded.	^{253}Es and ^{255}Fm both formed by multiple neutron capture by ^{238}U, yielding ^{253}U and ^{255}U. These decay by a series of β-emissions to give the Es and Fm, which then decay by α-emission.	—

to the intense radioactivity of the shorter lived isotopes with the consequent health hazard and heating and chemical effects on the solvent (see question 4.1). The use of glove boxes became mandatory and remote handling techniques had to be developed to deal with quantities above the 100 μg scale.

4.2. *Elements* 97–100

4.2.1. *Synthesis*. These elements are synthesized in principle by two processes: (1) multiple neutron capture with subsequent β-decay of neutron rich isotopes; (2) bombardment with accelerated charged particles. The first of these processes is alone capable of producing macroscopic amounts.

(1) Multiple neutron capture can be carried out by (*a*) prolonged exposure to a high and steady neutron flux in a High Flux Isotope Reactor (p. 43), or (*b*) by a very high, but very short pulse in a thermonuclear explosion.

(*a*) A quantity of ^{239}Pu is first irradiated for about eighteen months in a normal reactor. This produces about 10% of ^{242}Pu, ^{243}Am, and ^{244}Cm with about 90% of fission products. Complete burn-up of the ^{239}Pu is essential because of the high proportion of fissions which occur with ^{239}Pu and ^{241}Pu and the consequent energy release. This energy cannot be easily dissipated in the High Flux Reactor in which the ^{242}Pu, ^{243}Am and ^{244}Cm, after separation, are irradiated for a further eighteen months. This second irradiation produces curium, berkelium and californium in decreasing amounts. A typical charge of 1 kg of ^{239}Pu produces about 3 g of californium mainly as ^{252}Cf which has a long half-life. The californium may be separated and again irradiated in the High Flux Reactor to produce a few μg of einsteinium and an even smaller quantity of fermium. The high losses in this step are due to fission (Figure 4.1).

(*b*) The explosion is produced by surrounding a nuclear detonator (^{235}U or ^{239}Pu) with LiD and LiT and enclosing the whole in a blanket of uranium. The detonator raises the temperature to $\approx 10^8$ K causing the fusion reaction

$$D + T \rightarrow He + n \tag{4.1}$$

to occur which raises the temperature still further. The high energy neutrons released on fusion are reduced to resonance energies in about 10^{-13} s and then react with the uranium shield by multiple neutron capture yielding isotopes very rich in neutrons with mass numbers up to 257. The whole process takes about 10^{-7} s. Successive $\beta-$decays produce the heavy elements (Figure 4.2). The yields of einsteinium and fermium are small but measurable.

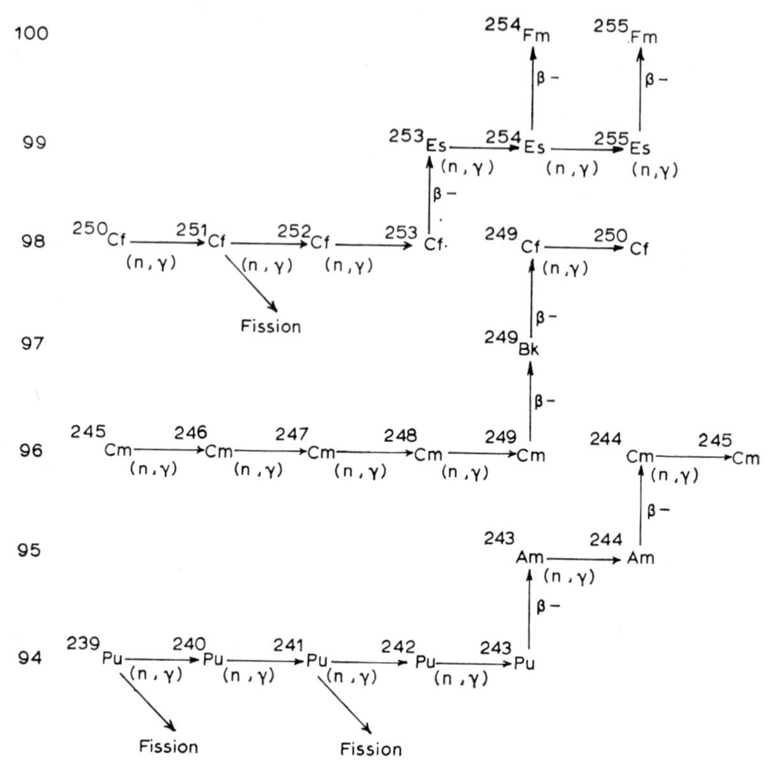

Figure 4.1. Formation of transplutonium elements by multiple neutron capture. ^{239}Pu is irradiated for about eighteen months in a normal reactor, producing about 10% of ^{242}Pu, ^{243}Am and ^{244}Cm with about 90% of fission products. The ^{242}Pu, ^{243}Am and ^{244}Cm are then separated and irradiated in a High Flux reactor (p. 43) for a further eighteen months, forming isotopes of Cm, Bk and Cf, in decreasing amounts. The Cf may be separated and again irradiated in the High Flux reactor to produce μg quantities of Es and Fm.

Taking the yield of Pu, Am, Cm, of $A \approx 245$ as 100 the corresponding yield of einsteinium and fermium would be $\approx 4 \times 10^{-4}$. The highest mass number so far obtained is 257. Beyond this number neutron binding energies rapidly decrease and fission processes break the neutron capture chains almost quantitatively.

(2) Bombardment with accelerated charged particles. The energy of the particle must exceed the coulomb barrier.† The compound nucleus produced has an excitation energy of about half the energy of the charged particle. This energy is rapidly lost

† i.e. the force of repulsion exerted by the positive charge of the nucleus.

either by fission or by neutron emission. When heavy ions such as ^{22}Ne are used on a uranium nucleus about one atom in 10^7 of the product undergoes neutron emission rather than fission and the yields are very small. When smaller charged particles, such as α-particles are used the coulomb barrier is much lower and they do not need to be so highly energetic. The compound nucleus formed has in consequence a much lower excitation energy so that neutron emission rather than fission is more likely. However the production of elements beyond $Z=102$ is impossible at present by the use of α-particles because targets of $Z>100$ of sufficient size are not available.

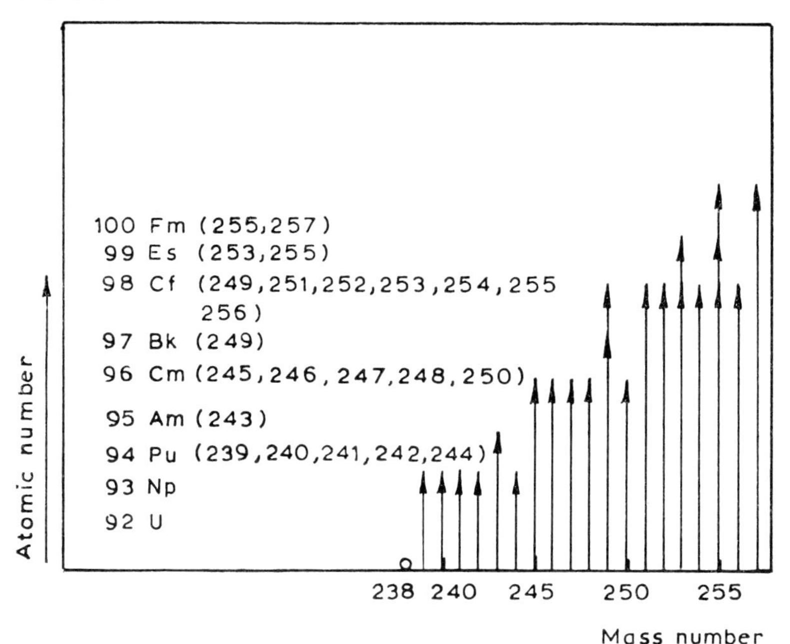

Figure 4.2. Production of heavy nuclides by means of a thermonuclear explosion. So far the heaviest nuclide produced by this method is $^{257}_{100}Fm$. A proposal has been made that isotopes of Es and Fm, produced by a thermonuclear explosion, should be used as targets of timed nuclear explosions. This would allow fast β-decay between the intense neutron irradiations and circumvent the high fission probability of even-even nuclei. The technical problems are considerable. (Reproduced by permission, and modified, from *Man-Made Transuranium Elements* by G. T. Seaborg, Prentice-Hall Inc., Englewood Cliffs, N.Y., 1963.)

4.2.2. *Identification.* In order to identify these elements it was necessary to develop the ion exchange technique using new eluants and to make use of the similarity between the eluting positions of the actinides and the corresponding lanthanides. The elements are eluted in order of decreasing atomic number. The target elements used were intensely radioactive and complicated remote control apparatus had to be designed to protect the operators.

Concentrated hydrochloric acid proved very useful for bulk separations of actinides from fission products, but this gave poor resolution of the heavy actinides. One of the most useful eluants was found to be ammonium 2-hydroxy-2-methylpropanoate (α-hydroxyisobutyrate). Figure 4.3 on p. 59 shows the order in which the actinides are separated.

The chemical properties of these elements were first studied on the tracer scale, but subsequently it has been possible to confirm the early work with macroscopic amounts of elements 97, 98 and 99. These studies revealed that they are typical actinides.

4.3. *Element* 101

Elements beyond fermium become increasingly difficult to prepare by multiple neutron capture, owing to the spontaneous fission of ^{256}Fm and ^{258}Fm. At present no β-active isotopes of fermium are known. It was therefore decided to bombard ^{253}Es with accelerated α-particles. The possible yield at saturation, where the rate of production equals the rate of decay, could be calculated from the equation:

$$N = N_A \sigma I t \tag{4.2}$$

where N = the number of atoms of 101 produced, N_A = the number of target atoms, σ = the capture cross-section, I = the intensity of the α-particle beam and t = the time of irradiation (see p. 74, Equation (5.15), which gives the saturation activity produced by irradiation of a target in a known neutron flux. (The activity multiplied by the time gives the number of atoms produced.)

The number of atoms of ^{253}Es available for this experiment was 10^9, σ was estimated at 10^{-3}b(10^{-24} cm^2), I was 10^{14} cm^{-2}s^{-1} and t was 10^4s. Hence $N = 10^9 \times 10^{27} \times 10^{14} \times 10^4 = 1$: i.e. the expected yield was one atom per experiment.

The einsteinium was electroplated on a thin gold foil which was placed in the path of the α-particle beam. A second 'catcher' foil was placed immediately behind the first so that any product would recoil on to the second foil. The 'catcher' gold foil was dissolved in 'aqua regia' and placed on a cation column. One spontaneous fission was observed in the eluant which came off the column, in the position expected for element 101. The experiment was repeated a number of times and thirteen spontaneous fissions were recorded.

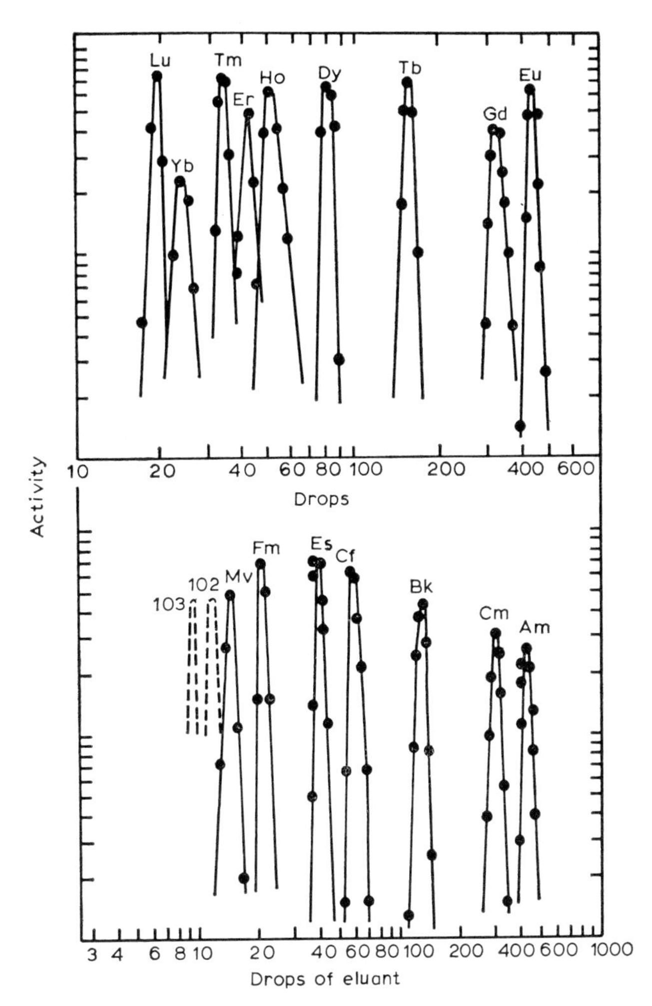

Figure 4.3. Elution of homologous actinides and lanthanides from Dowex-50 at 87°C, with 2-hydroxy-2-methylpropanoate as eluant. The elements are eluted in order of decreasing atomic number. This was the principal method used for identifying the heavier actinides. (Reproduced by permission, from *Man-Made Transuranium Elements* by G. T. Seaborg, Prentice-Hall Inc., Englewood Cliffs, N.Y., 1963.)

five in the eluant from the expected position of 101 and eight in the eluant from the previously determined position for element No. 100. It thus became clear that $^{256}101$ had been formed and had decayed by electron capture to ^{256}Fm which was known to decay by spontaneous fission with a half-life of 3·5 h [27]. Element 101 was named mendelevium, after the founder of the periodic table.

Mendeleef would have been astonished at a technique which could identify one single atom of a new element, but without the help of his periodic table this identification would have been impossible.

Subsequent experiments using larger targets and heavier isotopes of einsteinium have produced additional isotopes of mendelevium, in particular ^{258}Md with a half-life of two months. Tracer studies with this isotope have shown that, in addition to the typical +3 oxidation state mendelevium also has a +2 state. In this it shows a resemblance to its lanthanide analogue terbium.

4.4. *Element* 102

The first reported production of this element was in 1957 from the Nobel Institute in Stockholm. ^{244}Cm was bombarded with $^{13}C^{4+}$ ions and an isotope of element 102 was claimed to have been formed

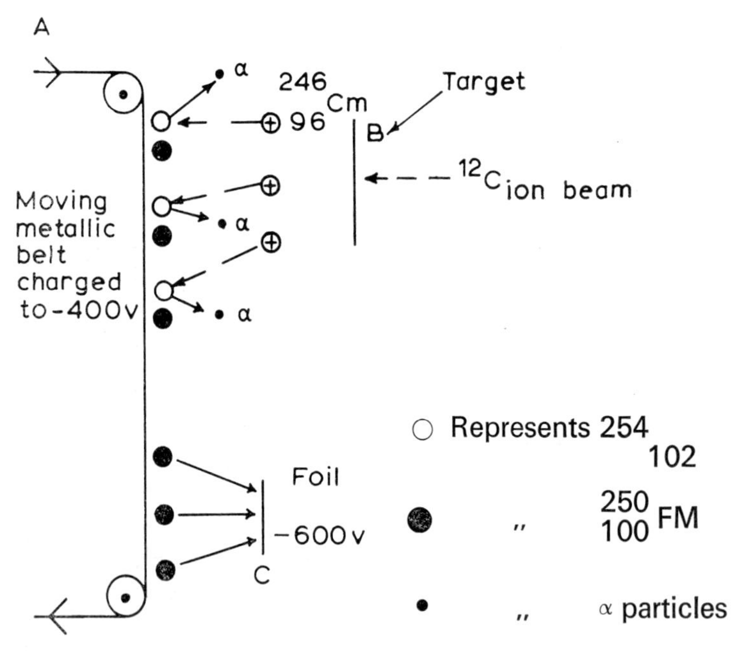

Figure 4.4. The production of element 102 (nobelium). A moving metallic belt A was placed in front of a target B, containing ^{246}Cm, which was bombarded with ^{12}C ions. A few atoms of ^{254}No were formed and knocked out of the target. These atoms, being positively charged, were collected on the belt. They decayed very rapidly to form ^{250}Fm, which was collected on the catcher foil C. Nobelium was thus identified from its daughter product of known characteristics.

$$^{246}_{96}Cm + ^{12}_{6}C \rightarrow ^{254}_{102}No + 4^{1}_{0}n \, . \quad ^{254}_{102}No \rightarrow ^{250}_{100}Fm + ^{4}_{2}He$$

with a half-life of about ten minutes. The element was named nobelium after Alfred Nobel. However attempts to confirm this result have failed. A year later, at Berkeley (USA) definite evidence for the formation of nobelium was obtained. A target of ^{246}Cm mixed with other isotopes was bombarded with ^{12}C ions. A moving metallic belt, charged negatively, was arranged directly behind the target (Figure 4.4). The atoms of nobelium, which were positively charged at the moment of formation, were attracted to the negatively charged belt. A piece of gold foil charged negatively with respect to the belt was fixed a short distance below the target. ^{250}Fm was identified on the foil, by its half-life of 38 s, by its elution position from an ion exchange column and from the known energy of the α-particles given off on decay. This indicated clearly that ^{252}No had been formed and had decayed according to the scheme:

$$^{254}_{102}No \rightarrow ^{250}_{100}Fm + ^{4}_{2}He \tag{4.3}$$

The experiment was repeated and the moving belt was led in front of a series of alpha detectors (Figure 4.5). This enabled a direct measurement of the half-life to be made (3 s).

Thus element 102 was identified chemically through its daughter product the properties of which were already known. Recent work has produced additional isotopes with longer half-lives, up to 3 min. With the aid of the 3 min activity it has been shown that nobelium has two oxidation states +2 and +3 and that the +2 state is the more stable. In this there is a close resemblance to ytterbium, the lanthanide analogue [28].

4.5. *Element* 103

This element, the last of the actinides, was prepared in 1961 at Berkeley by bombarding a californium target with ^{10}B and ^{11}B ions. The experimental arrangement was similar to that shown in Figure 4.5 and an isotope emitting α-particles of 8·6 MeV energy with a half-life of 4 s was detected. The element was named lawrencium after E. O. Lawrence, the inventor of the cyclotron [28].

Further work by Russian scientists has produced ^{256}Lr with a half-life of 25 s. Using eight atoms of this isotope lawrencium was extracted from aqueous solution into an organic solvent and the partition coefficient determined over the pH range 2–3 [29]. Its behaviour was identical to Cf and Fm thus indicating that the stable state in aqueous solution was the +3 ion. No evidence for a +2 state was found. This compares with lutecium, the final member of the lanthanide series.

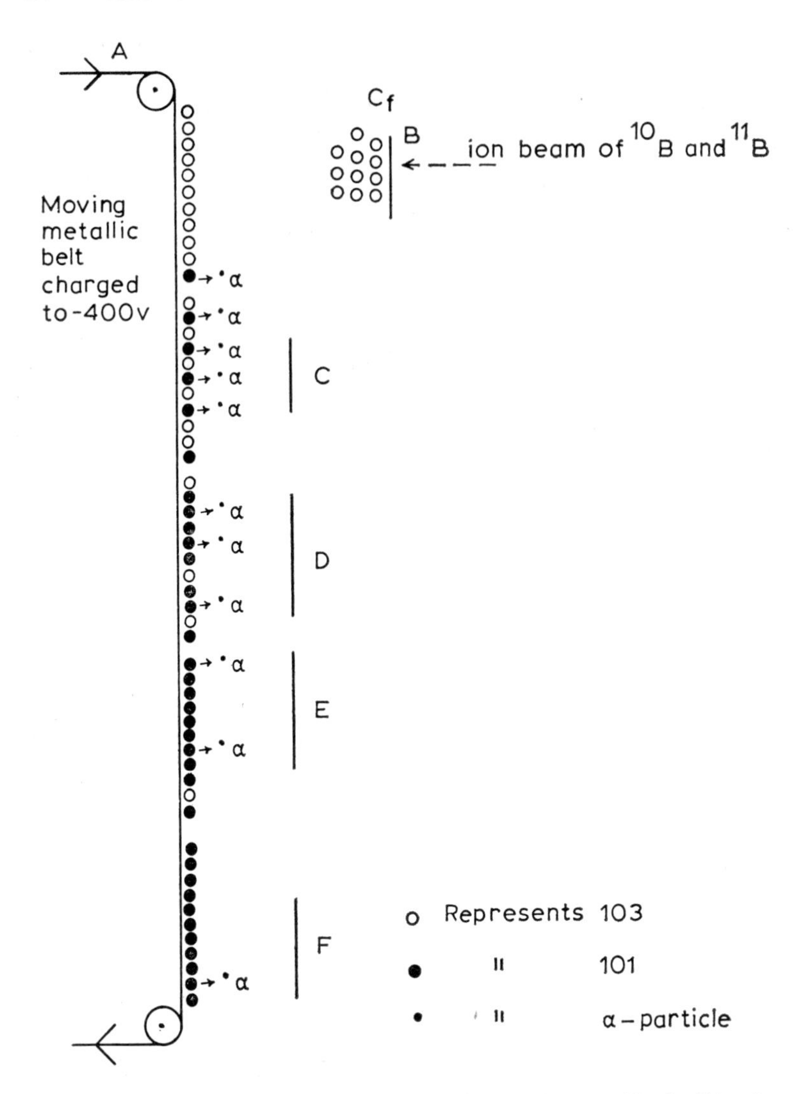

Figure 4.5. Production and collection of ^{257}Lr. A target B of californium was bombarded with ^{10}B ions. A moving belt A, charged negatively, collected the product. The belt passed in front of a series of α-detectors C, D, E, F. From the known rate of movement of the belt and the number of counts recorded by each of the α-detectors the half-life of the product was determined as 4s. In addition the energy of the α-particles was measured as 8·6 MeV. No such nuclide was known, indicating that the Cf and B nuclei had fused to form element 103 which was named lawrencium.

$$^{252}_{98}\mathrm{Cf} + ^{10}_{5}\mathrm{B} \rightarrow ^{257}_{103}\mathrm{Lr} + 5^{1}_{0}\mathrm{n}$$

4.6. *Transactinide elements*

The production of lawrencium completed the actinide series. The next element was expected to start another transition or ' d-block ' series. Element 104 should resemble hafnium and 105 tungsten. A Russian claim to the preparation of 104 was made in 1964. They bombarded ^{244}Pu with ^{22}Ne ions. Two activities were detected with half-lives of 0·1 and 4·5 s respectively. Both decayed by spontaneous fission which is not a good method of identification [29]. The Berkeley group in the United States obtained more positive evidence in 1969 by bombarding ^{249}Cf with ^{12}C and ^{13}C ions. Two alpha-emitting isotopes were obtained with half-lives of 4·5 and 3 s respectively. The alpha energies of both these isotopes were determined, and the daughter products identified as known isotopes of nobelium. More recently a further isotope $^{261}104$ has been prepared with a half-life of 65 s [30].

Chemical evidence using isotopes of such short half-lives is difficult to obtain, but the 65 s activity has been eluted from a cation exchange column along with zirconium and hafnium and apart from the actinides. The Russians have detected a volatile chloride similar to the chlorides of zirconium and hafnium [31]. This is good evidence for a transactinide element. The name of 104 has not yet been settled. The Russians have called it kurchatovium and the Americans rutherfordium. There seems no reason to doubt the production of 104.

4.7. *Element* 105

Again both Russian and American claims have been made for the first synthesis of this element in 1969 and 1970 respectively [32], [33]. The American claim is the more convincing. ^{249}Cf was bombarded with ^{15}N ions and an alpha activity of 9·1 MeV energy was observed which decayed with a half-life of 1·6 s. The daughter produced was identified as ^{256}Lr from its known half-life and alpha energy, thus proving that the parent must have been $^{260}105$. In addition a $^{262}105$ isotope was produced with a half-life of 40 s. This isotope should facilitate chemical tracer studies [34]. The americans have proposed the name hahnium (Hn).

4.8. *Super-heavy elements*

The production of elements of higher atomic number than 105 becomes increasingly difficult owing to nuclear instability. The principle difficulties are first to attain a sufficiently high neutron to proton ratio and secondly to produce a compound nucleus with a low excitation energy in order to avoid spontaneous fission. Such a nucleus should have the special groupings of nucleons known as 'magic numbers'.

4.9. *Magic numbers*

The electrons surrounding the nucleus of the atom form certain very stable groups. The numbers in these groups are 2, 8, 18, 32, i.e. twice the squares of the natural numbers. In the same way there are certain stable groups of nucleons which can be seen from studying the distribution and abundance of the stable isotopes. The numbers are referred to as 'Magic Numbers' because there is as yet no completely satisfactory explanation of their occurrence.

It is an interesting fact that neutrons and protons favour even-numbered combinations, that is they show a strong tendency to pair, like electrons with opposed spins. The elements with even atomic numbers have far more stable isotopes than those with odd atomic numbers. Stable isotopes with odd numbers of protons and odd numbers of neutrons are very rare indeed. The following table shows the numbers of known stable nuclei against the even or odd numbers of neutrons and protons.

Number of stable nuclei	Number of neutrons (N)	Number of protons (Z)
163	even	even
55	even	odd
50	odd	even
4	odd	odd

Certain even-numbered arrangements of protons or neutrons are particularly stable. The first such number is 2; ${}^{4}_{2}He$ has a particularly high binding energy. The next number is 8, ${}^{16}_{8}O$, again with an abnormally high binding energy; oxygen is the most abundant element in the earth's crust. The next number is 20; ${}^{40}_{20}Ca$ is the most abundant of calcium's six isotopes (96%) and also the lightest; all stable isotopes heavier than ${}^{40}Ca$ have more neutrons than protons. The next number is 28; Ni with 28 protons has five stable isotopes; ${}^{52}_{24}Cr$ with 28 neutrons has an abundance of 83% and ${}^{51}_{23}V$, also with 28 neutrons, has an abundance of 99%.

The next number 50 is a very obvious magic number. Tin with $Z = 50$ has no less than ten stable isotopes, more than any other element, and it shows a higher abundance in the Earth's crust than neighbouring elements. Examples of stable isotopes with 50 neutrons are $^{90}_{40}Zr$ with an abundance of 51% and the lightest of the zirconium isotopes, $^{89}_{39}Y$, with an abundance of 100%, and $^{88}_{38}Sr$ with an abundance of 82% and the highest N/Z ratio of the strontium isotopes.

Perhaps the most striking magic number is 82. Lead with $Z=82$ has four stable isotopes. It is the end product of the three natural decay series. There are no less than seven stable nuclei with 82 neutrons, $^{136}_{54}Xe$ with an abundance of 8·9%, $^{138}_{56}Ba$ with an abundance of 72%, $^{139}_{57}La$ with an abundance of 99·9%, $^{140}_{58}Ce$ with an abundance of 88%, $^{141}_{59}Pr$ with an abundance of 100%, $^{142}_{60}Nd$ the most abundant of the neodymium isotopes and also the lightest, and finally $^{143}_{61}Sm$, the lightest of the stable samarium isotopes, the next two being unstable.

The last natural magic number is 126; $^{208}_{82}Pb$ has both 126 neutrons and 82 protons and is the most abundant stable isotope of lead (52%). It is the end product of the thorium decay series. $^{209}_{83}Bi$ with 126 neutrons is the heaviest stable isotope known and the only stable isotope of bismuth.

Theoretical calculations have been made which indicate that $Z=114$ and $N=184$ may also be magic numbers, though these calculations are somewhat uncertain [35]. A nuclide $^{298}114$ should be doubly magic, like lead, and should be stable to spontaneous fission and also to α- and β-decay. An element with $Z=114$ should be a homologue of lead and searches are being made in lead ores for the existence of natural isotopes of this element.

Considering possible methods of synthesis, bombardment of neutron rich isotopes with heavy ions is unlikely to produce a sufficiently high value of A. The following reaction has been attempted:

$$^{248}_{96}Cm + ^{40}_{18}Ar \rightarrow ^{284}114 + 4^{1}_{0}n \tag{4.4}$$

The argon ions were accelerated to 250 MeV, but no success was reported. The value of A was clearly too low, well below the beta stability line and in the region which is highly unstable to spontaneous fission and alpha decay. The compound nucleus would have been formed in a highly excited state.

Various other methods have been suggested, such as using a target and projectile with Z around 50 to 65 and controlling the projectile energy so that the resulting compound nucleus would have a low excitation energy. One such possible reaction would be:

$$^{161}_{64}Gd + ^{132}_{50}Sn \rightarrow ^{293}114$$

The experimental difficulties are considerable as both target and projectile have half-lives of a few minutes only. Another more hopeful method which has been proposed is to fuse two heavy nuclei and hope that the resultant very large mass will undergo highly unsymmetrical fission. This suggestion is based on the observation that when heavy nuclei are bombarded with projectiles of increasing mass the maximum in the fission yield moves to higher mass numbers.

A novel method of synthesis was attempted in 1971. A tungsten target was bombarded with 24 000 MeV protons for a period of several months. Elastic scattering of the protons gave some tungsten nuclei recoil energies in excess of 1000 MeV, which is the value of the coulomb barrier between two tungsten nuclei. Collision of one of these nuclei with another tungsten nucleus could produce fusion of the two nuclei with probable highly unsymmetric fission. This might produce a superheavy nucleus. A very thorough analysis of the tungsten target was made using carriers of all the elements from osmium to lead. A very low activity was detected in the mercury fraction indicating the possibility of the production of element 112, but the results so far are inconclusive [36].

It has only been possible here to indicate possible lines of advance which might be used. Special accelerators capable of producing beams of heavy particles are now under construction in several countries and the results obtained by use of these machines must be awaited.

Question

4.1. Curium-242 has a half-life of 162·5 days and decays by α-emission. The mean energy of the α-particles is 6 MeV. If you had 10 μg of this isotope and wished to carry out ultramicrochemical studies in 0·1 M solution calculate the time it would take for your solution to boil starting at 20 °C. (Assume 50 % heat losses.)

5. The laws of radioactivity

5.1. *Units*.

The traditional unit of radioactivity is the 'curie' (Ci). This was originally defined as the rate of disintegration of radon in equilibrium with 1 g of radium, i.e. $3{\cdot}7 \times 10^{10}$ disintegrations s^{-1}. The unit has now been extended to cover any radioactive species which decays at this rate. The curie is an unnecessarily large unit. Sub-units are the millicurie, mCi ($3{\cdot}7 \times 10^{7}$ dis s^{-1}) and the microcurie, μCi ($3{\cdot}7 \times 10^{4}$ dis s^{-1}). A new unit (S.I.), the 'becquerel' (dis s^{-1}) has been suggested, but at present is rarely used.

5.2. *Decay*

Reference to this has already been made (p. 4), but for the sake of completeness the theory will be briefly discussed here. The rate of decay of a single radioactive species follows the first order law, familiar to students of elementary reaction kinetics:

$$-\frac{\mathrm{d}N}{\mathrm{d}t} = \lambda N \qquad (5.1)$$

where N is the number of radioactive atoms present and λ is the disintegration constant, similar to a velocity constant. It differs from a velocity constant, however, in that it is not temperature dependent. Furthermore, radioactive decay is a true 'unimolecular' (uniatomic) process, unlike most first order chemical reactions which must be at least bimolecular since chemical reaction only proceeds as a result of activation by collision. The rate of radioactive decay is entirely unaffected by external conditions.

Integration of Equation 5.1 gives:

$$\ln N = -\lambda t + G \qquad (5.2)$$

where G is the integration constant.

G may be evaluated by putting $t=0$ when N becomes N_0 and $G=\ln N_0$. Equation 5.2 may now be rewritten:

$$\ln\frac{N}{N_0}=-\lambda t \tag{5.3}$$

or:

$$N=N_0\exp(-\lambda t) \tag{5.4}$$

5.3. *Half-life*

Every radioactive isotope is characterized by its half-life (Experiments 4–7). Half-lives vary from 10^{10} y to 10^{-10} s. The half-life is the time taken for the number of radioactive atoms to decrease to half the initial value. Putting $N=\frac{1}{2}N_0$ and $t=t_{1/2}$ in Equation 5.3 and inverting:

$$\lambda t_{1/2}=\ln\frac{N_0}{\frac{1}{2}N_0}$$

$$=\ln 2$$

i.e.

$$t_{1/2}=0{\cdot}693/\lambda \tag{5.5}$$

Since $t_{1/2}$ is the time taken for the activity to fall to half of the initial value, so in the time $nt_{1/2}$ the activity falls to $(\frac{1}{2})^n$. Theoretically, the activity never becomes zero, but in ten half-lives it falls to about 1/1000, i.e. 0·1%, of the initial value.

5.4. *Average life*

The exponential law (5.4) indicates that there is a definite probability of any particular atom disintegrating at a given time. In fact, any particular atom may disintegrate at any time from $t=0$ to $t=\infty$, which explains the gradual process of decay. It is possible to calculate the average life of the atoms of a particular radioactive species. This average life $\tau=1/\lambda$ (for a proof of this see Appendix 1).

Since

$$t_{1/2}=0{\cdot}693/\lambda \tag{5.5}$$

$$\tau=t_{1/2}/0{\cdot}693. \tag{5.6}$$

τ is the time taken for the activity to fall to 1/e of the initial value.

5.5. *Growth of a radioactive product*

When a radioactive nuclide forms a daughter product which itself is radioactive, the rate of growth of the daughter activity is determined by the difference between the rate of decay of the parent and the rate of decay of the daughter. Consider a system such as:

$$\mathrm{A}\rightarrow\mathrm{B}\rightarrow\mathrm{C}\rightarrow\mathrm{D}$$

The rate of decay of A by Equation (5.1) $=\lambda_A N_A$, and similarly the rate of decay of B, independently of A, $=\lambda_B N_B$. The rate of growth of B in the presence of A is:

$$\frac{dN_B}{dt}=\lambda_A N_A-\lambda_B N_B \tag{5.7}$$

but $N_A=N_{0,A}\exp(-\lambda_A t)$ (Equation (5.4)).
Re-writing (5.7) and substituting for N_A:

$$\frac{dN_B}{dt}+\lambda_B N_B-\lambda_A N_{0,A}\exp(-\lambda_A t)=0 \tag{5.7 a}$$

The solution of this equation (see Appendix 2) gives:

$$N_B=\frac{\lambda_A}{\lambda_B-\lambda_A}N_{0,A}[\exp(-\lambda_A t)-\exp(-\lambda_B t)]+N_{0,B}\exp(-\lambda_B t) \tag{5.8}$$

5.6. *Radioactive equilibrium*

This is defined as the state of the system when the various daughter products are growing and decaying at the same rate, and the activity of the system has reached a steady value. There are two types of radioactive equilibrium—secular equilibrium and transient equilibrium.

5.6.1. *Secular equilibrium.* If the rate of decay of the parent A is very slow compared to that of the daughter B, i.e. $\lambda_A \leqslant \lambda_B$, B can be considered to form at a constant rate from A.

Applying this to the general equation (5.8), λ_A can be neglected in comparison to λ_B and $\lambda_A t$ is effectively $=0$, i.e. $\exp(-\lambda_A t)=1$. Putting $N_{0,B}=0$ at $t=0$, the last term vanishes and the equation reduces to:

$$N_B=\frac{\lambda_A}{\lambda_B}N_{0,A}(1-(\exp-\lambda_B t)) \tag{5.9}$$

N_B reaches a maximum value when the exponential term vanishes. It is instructive to calculate the proportion of this maximum reached for different values of t in terms of the half-life of B. When $t=t_{1/2}$, $\lambda_B t=0{\cdot}693$ (since $\lambda=0{\cdot}693/t_{1/2}$) (Equation (5.5)) and $\exp(-\lambda_B t)=0{\cdot}5$. Thus N_B reaches half of its maximum value after one half-life. When $t=2t_{1/2}$, $\exp(-\lambda_B t)=0{\cdot}25$ and N_B reaches three-quarters of the maximum. When $t=7t_{1/2}$, $\exp(-\lambda_B t)=0{\cdot}01$ and N_B reaches 99% of the maximum, and when $t=10t_{1/2}$, N_B has reached 99·9% of the maximum. For most purposes N_B can be considered to have reached a steady state after seven half-lives. This state is called secular equilibrium (Figure 5.1).

Since $1-\exp(-\lambda_B t)=1$ when equilibrium has been attained,

$$\lambda_A N_{0,A}=\lambda_B N_B.$$

These conditions prevail in a radioactive mineral such as pitchblende where the long chain of decay products are all decaying at the rate of the parent, i.e. $\lambda_A N_A=\lambda_B N_B=\lambda_C N_C=\lambda_D N_D$ etc.

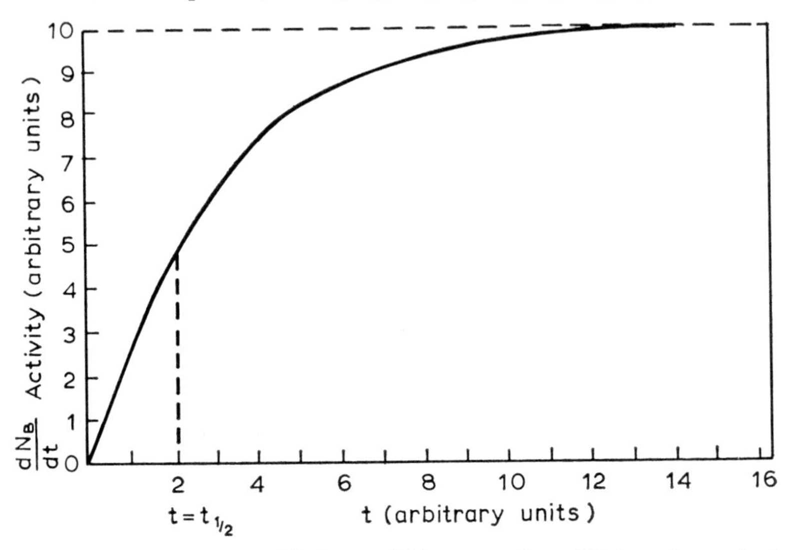

Figure 5.1. Secular equilibrium. This type of equilibrium is reached when a daughter product decays at the same rate as it is being formed from a parent with a long half-life. The figure represents the growth of a daughter product from the parent after initial separation. Secular equilibrium may be considered to be reached, for all practical purposes, after seven half-lives of the daughter. Half of the maximum activity is reached after one half-life.

5.6.2. *Transient equilibrium.* When the decay of the parent takes place at a measurable rate and the half-life of the parent is greater than that of the daughter by about one order of magnitude, the combined activity of the system, with the daughter separated initially, grows to a maximum and then decays, finally reaching transient equilibrium when the system decays with the half-life of the parent (Figure 5.2). This is called transient equilibrium (Experiment 17). Strictly, any system in which the half-life of the parent is greater than that of the daughter should show transient equilibrium, but in many cases the rate of decay of the parent is too slow to show measurable decay.

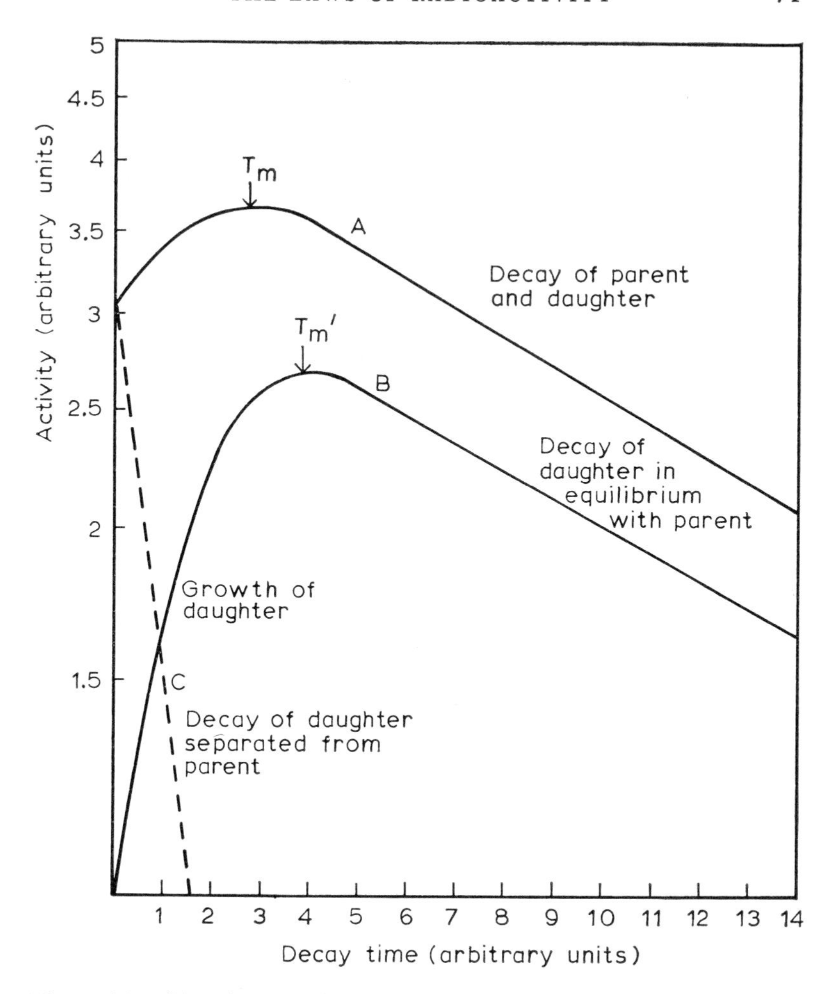

Figure 5.2. Transient equilibrium. This occurs when the half-life of the daughter is less than that of the parent and the parent shows a measurable rate of decay. Curves A and B show the activity of a sample of the parent from which the daughter has been separated. In A the emissions of both parent and daughter have been counted, whereas in B only the emissions of the daughter have been counted. C shows the decay of the daughter separated from the parent (from measurements with ^{212}Pb and ^{212}Bi).

In the case of transient equilibrium Equation (5.8) reduces to:

$$N_B = \frac{\lambda_A}{\lambda_B - \lambda_A} N_{0,A} \exp(-\lambda_A t) \qquad (5.10)$$

since $\exp(-\lambda_B t)$ can generally be neglected in comparison to $\exp(-\lambda_A t)$ and $N_{0,B} = 0$ at $t = 0$.

Now:

$$N_A = N_{0,A} \exp(-\lambda_A t) \text{ (Equation 5.4).}$$

Hence, when transient equilibrium has been reached:

$$\frac{N_B}{N_A} = \frac{\lambda_A}{\lambda_B - \lambda_A} \tag{5.11}$$

5.7. *No equilibrium*

This comparatively unusual condition occurs when the half-life of the parent is shorter than that of the daughter. It is clear that in this case no equilibrium is possible, since Equation (5.10) gives N_B a negative value.

5.8. *Formation of artificial radionuclides by particle bombardment*

The production of radioactive isotopes by particle bombardment is of considerable practical importance. The particles are either positively charged such as protons, deuterons, alpha particles or other heavy ions, or fast or slow neutrons. Charged particles must be accelerated to high energies by powerful machines such as cyclotrons, linear accelerators or van de Graaf generators to enable them to penetrate the coulomb barrier surrounding the atomic nucleus. Fast neutrons can be produced by nuclear fission, or as secondary products as a result of bombarding beryllium targets with alpha particles, or tritium (^3H) targets, such as titanium tritide, with protons or deuterons. The compound nucleus formed as a result of capture of the bombarding particle is extremely unstable and emits another particle, or gamma ray.

The principal types of nuclear reactions which occur as a result of particle bombardment are as follows:

	Type of reaction	Relation of product to target elements	
		Mass number (A)	Atomic number (Z)
1	p,n	same	+1
2	n,p	same	−1
3	α,n	+3	+2
4	n,α	−3	−2
5	n,2n	−1	same
6	d,p	+1	same
7	n,γ	+1	same

Reaction 7 is brought about by slow neutrons, i.e. neutrons reduced to thermal energies (about 0·05 eV) in a reactor (p. 38), or with the aid of a laboratory neutron source surrounded by a suitable moderator (p. 141). This is the most important method used for producing radioactive isotopes.

During irradiation in a reactor both slow and fast neutrons are present, and they may yield different products from the same nuclide. Thus when sodium chloride is irradiated in a reactor the isotopes ^{24}Na, ^{36}Cl and ^{38}Cl are produced by capture of slow neutrons. In addition some ^{35}S and ^{32}P are produced by fast neutrons. The following equations represent these reactions:

	σ_a/barn
$^{23}_{11}Na + ^{1}_{0}n \rightarrow ^{24}_{11}Na$	0·53
$^{35}_{17}Cl + ^{1}_{0}n \rightarrow ^{36}_{17}Cl$	30
$^{37}_{17}Cl + ^{1}_{0}n \rightarrow ^{38}_{17}Cl$	0·6
$^{35}_{17}Cl + ^{1}_{0}n^* \rightarrow ^{35}_{16}S + ^{1}_{1}H$	0·19
$^{35}_{17}Cl + ^{1}_{0}n^* \rightarrow ^{32}_{15}P + ^{4}_{2}He$	0·1

where n* represents a fast neutron.

The number of 'events' per second, n, which occur as a result of particle bombardment is governed by the number of atoms of the target element (N_A), the cross-section of the target (σ) in barns, and the flux f (particles cm^{-2} s^{-1}). This assumes that the target is thin and that the flux is not appreciably reduced by capture,

i.e.
$$n = N_A \sigma f \tag{5.12}$$

These 'events' may be neutron capture, nuclear disintegration, or nuclear fission.

Equation (5.12) measures the rate of growth of a product in the absence of decay. Most products are radioactive, and the situation is similar to secular equilibrium where σf replaces λ_A of equation (5.9). Writing N_B to represent the number of atoms of radioactive product formed during time t secs and putting $N_{0,B} = 0$:

$$N_B = \frac{N_A \sigma f}{\lambda_B}[1 - \exp(-\lambda_B t)] \tag{5.13}$$

and substituting $-\dfrac{dN_B}{dt}$ for $\lambda_B N_B$:

$$-\frac{dN_B}{dt} = N_A \sigma f [1 - \exp(-\lambda_B t)] \tag{5.14}$$

Equation (5.14) gives the activity of the product after time t. Both N_B and $-dN_B/dt$ reach a maximum value when t is long compared to the half-life, and the exponential term becomes effectively $=0$ (see p. 70).

i.e.
$$-\frac{dN_B}{dt}=N_A\sigma f \qquad (5.15)$$

This maximum value is called the saturation value.

Equation (5.14) can be used to determine the yield of a radioisotope produced by irradiation in a given flux of slow neutrons for a given time. It can also be used to determine the neutron flux in an assembly by irradiating an element of known capture cross-section for a known time and measuring the activity produced. Gold foils or solutions of a gold salt are often used for this purpose owing to the high capture cross-section of gold for slow neutrons (98 b) (see Experiment 13).

If the time of irradiation is short compared to the half-life of the target, $[1-\exp(-\lambda_B t)]$ becomes effectively equal to $\lambda_B t$ and (5.14) simplifies to:

$$-\frac{dN_B}{dt}=N_A\sigma f\lambda_B t \qquad (5.16)$$

5.9. *Correction of disintegration rate for decay*

It is frequently necessary to correct an observed disintegration rate to find the rate of decay at some fixed time t_0.

Let N_0 be the number of atoms of the radioactive substance at t_0 and N_1 the number at the time of measurement t_1. Then the observed disintegration rate is:

$-\frac{dN_1}{dt}=\lambda N_1$ and the required disintegration rate is: $-\frac{dN_0}{dt}=\lambda N_0$.

Now: $N_1=N_0\exp(-\lambda t)$ (Equation 5.4)

Let: $\frac{dN_1}{dt}=I_1$, and $\frac{dN_0}{dt}=I_0$

Clearly: $\frac{N_1}{N_0}=\frac{I_1}{I_0}$

and we can write:

$$I_1=I_0\exp(-\lambda t).$$

Transposing and taking logs gives the required solution:

$$\ln I_0=\ln I_1+\lambda t \qquad (5.17)$$

5.10. *To determine an absolute decay rate*

For many purposes it is not necessary to determine the absolute decay rate I, provided the same measuring equipment is used throughout the experiment. Instead a quantity A is used which is the activity as measured with the instrument concerned, expressed as the number of counts recorded per unit of time. If the absolute decay rate must be known, as in the determination of neutron fluxes, it is necessary to calibrate the counter with a standard source of known disintegration rate. This source should be composed of the same isotope as the one being used in the measurement (see Experiment 6 for a more detailed treatment of this subject).

5.11. *The practical determination of half-lives*

The measurement of half-life is a valuable means of identification. The method used depends on the duration of the half-life. For nuclides with half-lives in the range of a minute to several years it is usual to plot the logarithm of the activity against the time and to determine the half-life from the slope of the graph (Figures 5.2 and 5.3). Measurements should be continued for several half-lives if possible. Ideally the same number of counts should be taken at each observation so that constant statistical accuracy may be maintained, but in practice this is frequently impossible. The time of counting should be recorded half way through the counting period. This assumes that the decay curve is uniform over this period. The assumption is justifiable provided that the counting period is not longer than half of the half-life. The error in this case is 0·5% (Experiments 4 and 5).

5.11.1. *Composite decay.* Sometimes radioactive specimens contain two or more nuclides which are decaying independently. Provided the half-lives differ by a factor of at least 2, it is relatively easy to resolve them. The plot of the logarithm of the activity against the time will start as a curve and eventually become a straight line when the activities of all but the longest lived nuclide have become negligible (Figure 5.3). To resolve the activities extrapolate the line DC backwards until it cuts the ordinate at E. From the slope of this line the half-life of the longer lived component may be determined. To determine the half-life of the shorter lived component, subtract the values of log activity at suitable time intervals along EC from the curve BC and plot the differences on the same graph. This will give the line FG from which the half-life of the shorter lived component may be calculated (Experiment 8). This principle may be extended to three or more components.

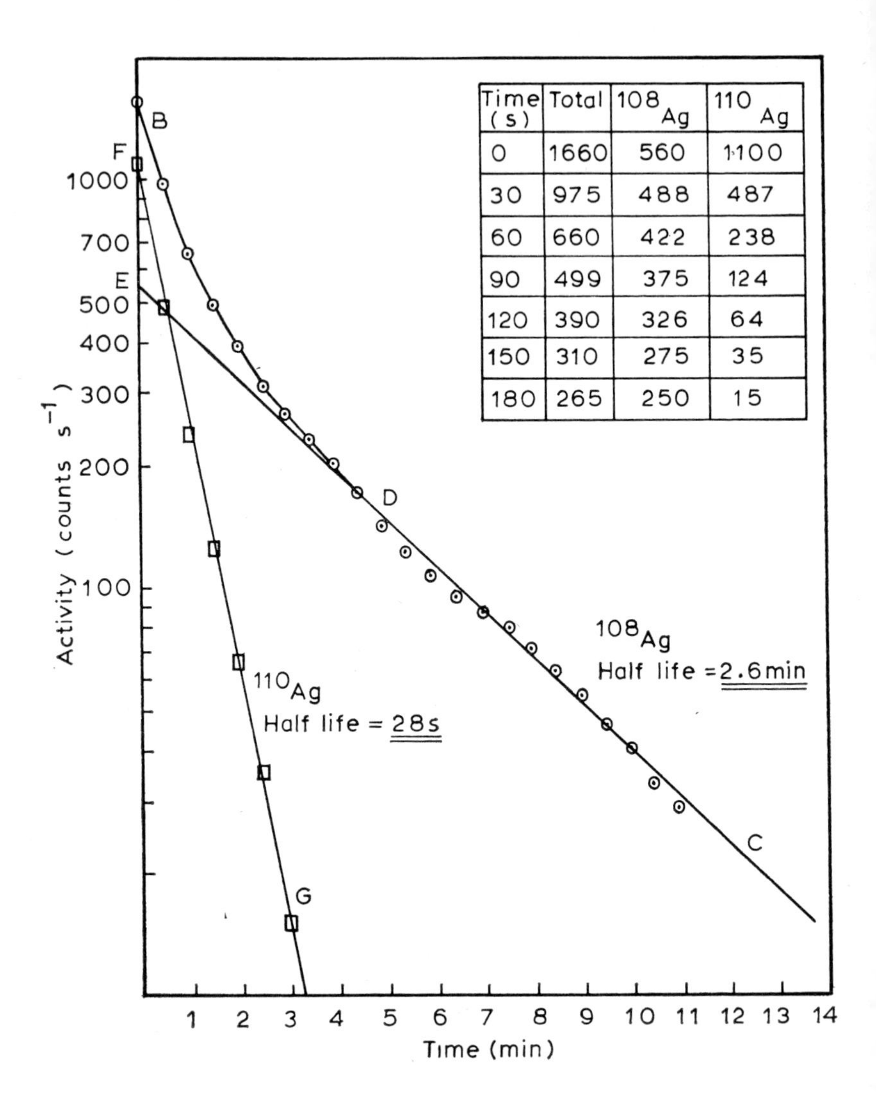

Time (s)	Total	^{108}Ag	^{110}Ag
0	1660	560	1100
30	975	488	487
60	660	422	238
90	499	375	124
120	390	326	64
150	310	275	35
180	265	250	15

Figure 5.3. Composite decay. When a sample contains two or more genetically unrelated nuclides, the decay graph, plotted on a log scale, starts as a curve, but ends as a straight line, when all but the longest lived nuclide have effectively decayed. The line FG is obtained by substracting ED from BD, and gives the decay graph of the shorter lived of two components. If the sample contains more than two nuclides the second graph will also start as a curve and must be resolved similarly. (The figure was drawn from measurements made in the author's laboratory; see Experiment 8.)

Better results may be obtained by suppressing some of the weaker activities by using an absorber such as aluminium foil.

5.11.2. *Very long half-lives.* When the half-lives are greater than a few years, decay measurements are not sufficiently accurate and other methods must be used. Provided the total number of radioactive atoms is known, the half-life can be determined from the absolute disintegration rate by use of the equation:

$$\frac{dN}{dt} = -\lambda N$$

To obtain the absolute disintegration rate it is necessary to know the detection coefficient, i.e. the efficiency of the counting instrument (Experiment 6). This method has been used to determine the half-life of ^{226}Ra (1622 y). It can also be used to determine the half-life of radioactive isotopes such as ^{40}K and ^{14}C mixed with inactive isotopes provided the proportion of the active isotopes is known. This proportion can be determined by the use of a mass spectrometer.

Another method depends on the establishment of radioactive equilibrium by use of the relationship:

$$\lambda_1 N_1 = \lambda_2 N_2 \quad \text{(p. 70)}$$

In this case the number of atoms of parent, N_1 and daughter, N_2 must be known, and also the half-life of the daughter element. The half-life of uranium and thorium can be obtained from the known half-lives of one of the decay products found in a uranium or thorium mineral, provided an accurate determination of the proportion of daughter to parent can be made. A good example is the determination of the half-life of uranium from the proportion of radium (^{226}Ra) to uranium found in pitchblende. The radium is the fifth member of the chain, but the relationship still holds because

$$\lambda_1 N_1 = \lambda_2 N_2 = \lambda_3 N_3 = \lambda_4 N_4 = \lambda_5 N_5 \text{ etc.}$$

The half-life of ^{226}Ra can be determined similarly by measuring the volume of ^{222}Rn in equilibrium with 1 g of ^{226}Ra since ^{222}Rn has a conveniently short half-life (3.8 d).

5.11.3. *Very short half-lives.* These are more difficult to determine. Neither of the above relationships can be used since it is not possible to determine the number of active atoms. One method depends on the use of a rapidly rotating disc. The sample is attached to the edge of the disc, which is rotated at constant speed. A counter such as a G-M tube is arranged to face the periphery. Knowing the diameter of the counter window the time for which the specimen is in front of the counter may be determined, and the counting interval from the speed of rotation. The number of counts each time the specimen passes in front of the counter window must be recorded automatically. The half-lives of short lived products in the gaseous state or in solution may be determined by the same principle, but using fluid flow. The gas or liquid flows down a tube at constant speed past a series of G-M tubes.

A method which is convenient for the determination of half-lives from 5 s to 10 min is the so-called 'ratio' method. Two measurements only are needed, the total number of counts over the whole period of the decay and the number from any given time to the end of the decay (Figure 5.4).

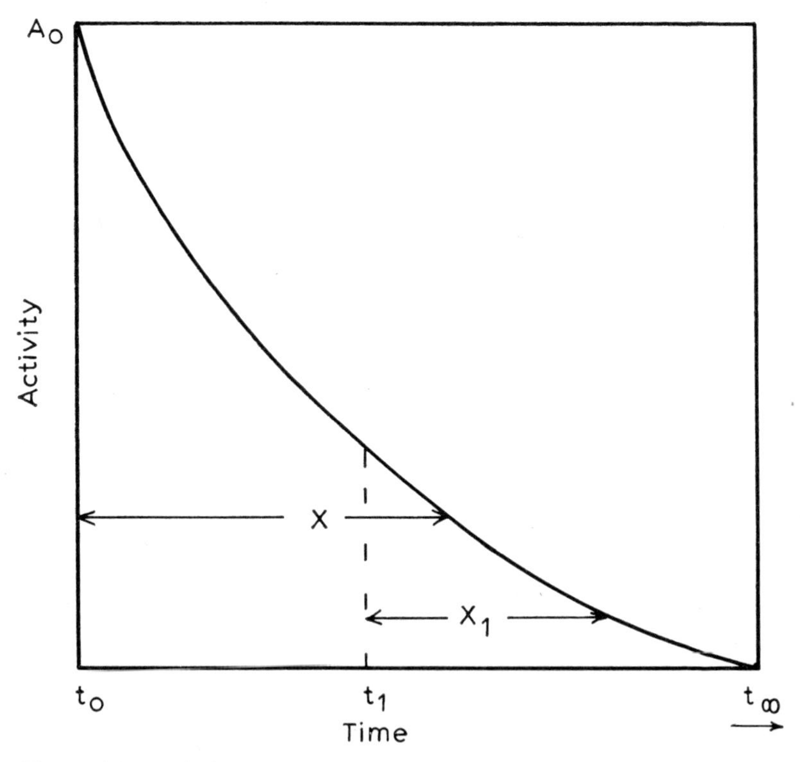

Figure 5.4. Half-life by the 'ratio' or 'two count' method. Two scalers are needed: one which records all counts over the whole period of the decay (x), and the other, which can be switched on at time t_1 (t_1 should be about twice the half-life) and gives the count x_1. Both scalers are switched off when the count rate falls to the background. The method is suitable for half-lives from a few seconds to a few minutes (see Experiment 7).

Let A_0 = the initial activity, which is proportional to the disintegration rate at t_0 and hence to the number of radioactive atoms present at t_0.

Let x = the total number of counts recorded over the whole period of the decay and x_1 = the number recorded between any time t_1 and the end of the decay. The number of counts is proportional to the number of radioactive atoms present and decaying.

Then:

$$x = \int_{t=0}^{t=\infty} A_0 \exp(-\lambda t)\mathrm{d}t = A_0/\lambda \tag{5.18}$$

and

$$x_1 = \int_{t=0}^{t=t_1} A_0 \exp(-\lambda t)\mathrm{d}t = A_0 [\exp(-\lambda t_1)]/\lambda \tag{5.19}$$

(See Appendix 3 for the solution of these integrations.)

Hence:

$$\frac{x_1}{x} = \exp(-\lambda t_1)$$

and:

$$\ln \frac{x_1}{x} = -\lambda t_1$$

$$= -\frac{0{\cdot}693 t_1}{t_{1/2}}$$

so:

$$t_{1/2} = \frac{0{\cdot}693 t_1}{\ln x - \ln x_1}$$

$$= \frac{0{\cdot}301 t_1}{\log_{10} x - \log_{10} x_1} \tag{5.20}$$

It can be shown that the most accurate result will be obtained when $x/x_1 = 5$, i.e. when t_1 is slightly greater than twice the half-life (Experiment 7).

For half-lives of less than 10 ms methods depending on separation are impracticable. The most satisfactory method is to use variable delayed coincidences between emissions from a relatively long lived parent and the very short lived daughter. The source S consisting of the parent in equilibrium with the daughter is arranged between two detectors D_1 and D_2 (Figure 5.5). The pulses from D_2 are delayed while those from D_1 enter the coincidence circuit directly. D_2 is sensitive to emissions from the parent only and D_1 from the daughter only. As the delay time is increased the number of coincidences increase to a maximum and then decrease (Figure 5.6). This maximum occurs at the average lifetime of the daughter. The half-life of the daughter may be determined from the average life by means of Equation 5.6.

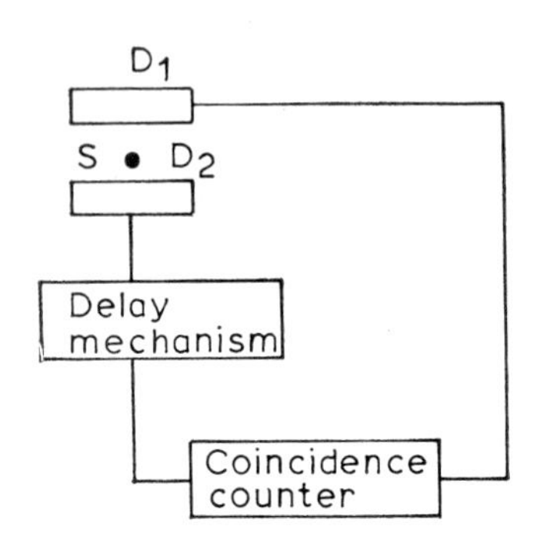

Fig. 5.5

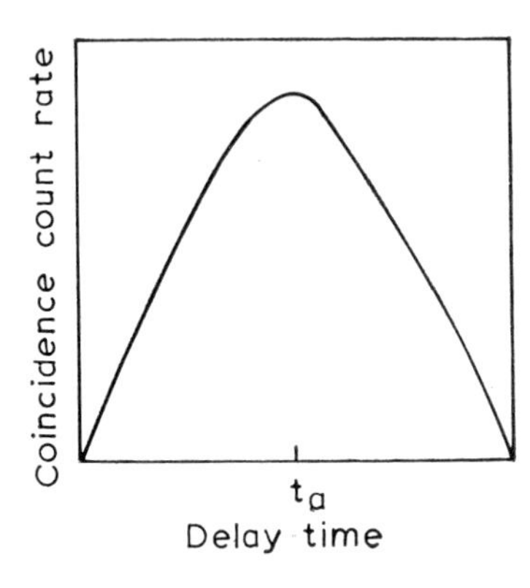

Fig. 5.6

Figures 5.5 and 5.6. Half-life by the method of delayed coincidences. The method is used for the measurement of half-lives so short that separation is impracticable. The source S, consisting of a relatively long lived parent in equilibrium with the daughter, is arranged between two detectors. D_2 is sensitive to emissions from the parent only and D_1 from the daughter only. A variable delay mechanism delays the counts from the parent. The coincidence count rate is plotted against the delay time. As the delay time is increased the coincidence count rate increases to a maximum and then decreases (Figure 5.6). This maximum occurs at the average life (p. 68) of the daughter (t_a). With scintillation counters it is possible to measure half-lives as short as 10^{-10}s.

If G–M tubes are used as detectors the method is limited to nuclides with half-lives of more than 50 μs owing to the relatively long resolving time of the tubes, but if scintillation counters or solid state detectors are used it is possible to measure half-lives as short as 10^{-10} s.

A general method which is applicable to both very long and very short-lived α-emitting nuclides, forming members of a radioactive series, depends on a relationship between the energy of the α-particles and the half-life. This relationship was discovered by

Geiger and Nuttall for the natural radioactive series, and was expressed in the form

$$\log\lambda = a + b\log R$$

where λ is the radioactive constant, R is the range (in cm of air) at STP, and a and b are constants. Since $\lambda = 0{\cdot}693/t_{1/2}$ and the range R is proportional to the energy, it is more convenient to plot $\log t_{1/2}$ against $\log E$ in MeV. The constant a varies from one series to another, but the constant b is the same for all. This relationship has been shown to hold for the artificial $(4n+1)$ series also. By measuring the half-lives of a few elements in the series and the energies of the α-particles emitted in each case it is possible, by use of the graph, to determine the half-lives of the remaining α-emitters from the energies of the α-particles. This method is not particularly accurate as the graph is not strictly linear over the complete range, but it gives a useful approximate value. In 1928 a theoretical treatment of the Geiger–Nuttall rule was made possible by the use of wave mechanics. The reader is referred to more advanced textbooks for this treatment. There is no simple relationship between the energies of β-particles and the half-life.

A few examples will now be given to illustrate the use of Equations (5.1)–(5.17).

Case 1. What is the mass in grams of 1 mCi of ^{131}I?

1 mCi has a disintegration rate of $3{\cdot}700 \times 10^7$ disintegrations s^{-1}. $t_{1/2}$ for $^{131}I = 8{\cdot}02$ d.

First obtain the number of ^{131}I atoms. Equation (5.1) is used, ignoring the minus sign:

$$N = \frac{dN}{dt} \cdot \frac{1}{\lambda}$$

$$= \frac{dN}{dt} \cdot \frac{t_{1/2}}{0{\cdot}693} \quad \text{(Equation 5.5).}$$

Expressing $t_{1/2}$ in seconds gives:

$$N = \frac{3{\cdot}700 \times 10^7 \times 8{\cdot}02 \times 24 \times 60^2}{0{\cdot}693}$$

$$= 3{\cdot}70 \times 10^{13}$$

This has a mass of: $$\frac{3{\cdot}70 \times 10^{13} \times 131}{6{\cdot}02 \times 10^{23}}$$

$$= 8{\cdot}05 \times 10^{-9}\ \text{g}$$

Using one of the latest liquid scintillation counters (p. 111) it would be possible to determine $10^{-3}\,\mu$Ci with reasonable accuracy, i.e. $8{\cdot}05 \times 10^{-15}$ g of ^{131}I. This shows the extreme sensitivity of radiochemical techniques.

Case 2. What is the activity in disintegrations s^{-1} of ^{212}Bi contained in ^{212}Pb six hours after the ^{212}Pb had been separated from a quantity of a thorium mineral containing 232 g of thorium?

This is a case of transient equilibrium. At $t=0$, i.e. the moment of separation of the ^{212}Pb, radioactive equilibrium will exist in the thorium mineral provided no ^{220}Rn has escaped.

Let N_0 = number of atoms of thorium and N_A = the number of atoms of ^{212}Pb.

Then $N_A = \dfrac{\lambda_0 N_0}{\lambda_A}$ (p. 70).

Now $t_{1/2}$ for ^{232}Th $= 1{\cdot}39 \times 10^{10}$ y, $t_{1/2}$ for ^{212}Pb $= 10{\cdot}6$ h, $t_{1/2}$ for ^{212}Bi $= 1$ h and 232 g of Th contains $6{\cdot}02 \times 10^{23}$ atoms.

Remembering that $\lambda = \dfrac{0{\cdot}693}{t_{1/2}}$ (Equation 5·5)

$$N_A = \frac{10{\cdot}6 \times 6{\cdot}02 \times 10^{23}}{1{\cdot}39 \times 10^{10} \times 365 \times 24}$$

$$= 5{\cdot}25 \times 10^{10} \text{ atoms}$$

Let N_B = number of atoms of ^{212}Bi after 6 h.

Initially $N_B = 0$, assuming a clean separation of ^{212}Pb. After 2.7 h the combined activity of the ^{212}Pb and ^{212}Bi will have reached the maximum. (This figure is obtained from Equation 5.8 by converting it to give the total activity and remembering that at $t=0$, $N_{0,B}=0$.) The total activity will be the sum of the activities of the ^{212}Bi and the ^{212}Pb, i.e.

$$= \lambda_B \frac{\lambda_A}{\lambda_B - \lambda_A} N_{0,A}[\exp(-\lambda_A t) - \exp(-\lambda_B t)] + \lambda_A N_{0,A} \exp(-\lambda_A t).$$

(See Equations 5.1 and 5.4.)

This expression must then be differentiated and equated to 0, (Appendix 4, p. 247).

After 6 h transient equilibrium will be effectively established. Hence determine the number of atoms of ^{212}Pb remaining; from this determine the number of atoms of ^{212}Bi in equilibrium using Equation 5.11, and finally the activity:

$$N_A = N_{0,A} \exp(-\lambda_A t) \quad \text{(Equation 5.4).}$$

$$\log_{10} N_A = \log_{10} N_{0,A} - \frac{\lambda_A t}{2 \cdot 3}$$

From this $N_A = 3{\cdot}54 \times 10^{10}$ atoms.

Now
$$N_B = \frac{N_A \lambda_A}{\lambda_B - \lambda_A} \quad \text{(Equation 5.11)}$$

$$= \frac{3{\cdot}54 \times 10^{10} \times 0{\cdot}065}{0{\cdot}693 - 0{\cdot}065}$$

$$= 3{\cdot}67 \times 10^9 \text{ atoms}$$

The activity of the ^{212}Bi:

$$-\frac{dN_B}{dt} = \lambda_B N_B \quad \text{(Equation 5.1)}$$

$$= \frac{0{\cdot}693 \times 3{\cdot}67 \times 10^9}{1 \times 60 \times 60}$$

$$= 7{\cdot}07 \times 10^5 \text{ dis s}^{-1}$$

Case 3. What disintegration rate would result from irradiating 1 g of sodium chloride in a flux of 10^{10} slow neutrons $\text{cm}^{-2}\,\text{s}^{-1}$ for 10 h?
Consider each nuclide in turn.

(*a*) ^{23}Na. $t_{1/2}$ for $^{24}\text{Na} = 15{\cdot}4$ h. The abundance of $^{23}\text{Na} = 100\%$, $\sigma = 0{\cdot}53$ b (p. 257). The time of irradiation is not short compared to the half-life of ^{24}Na, so that Equation 5.14 must be used.

$$-\frac{dN_B}{dt} = N_A \sigma f[(1 - \exp(-\lambda t)],$$

putting $-\dfrac{dN_B}{dt} = I_1, I_2$ and I_3 for ^{24}Na, ^{36}Cl and ^{38}Cl, respectively,

$$I_1 = \frac{6{\cdot}023 \times 10^{23}}{58{\cdot}5} \times 0{\cdot}53 \times 10^{-24} \times 10^{10}\left(1 - \exp - \frac{0{\cdot}693}{15{\cdot}4} \times 10\right),$$

$$= 1{\cdot}98 \times 10^7 \text{ dis s}^{-1} \text{ for } ^{24}\text{Na.}$$

(*b*) ^{35}Cl. $t_{1/2}$ for ^{36}Cl $= 3 \times 10^5$ y. The abundance of ^{35}Cl $= 75{\cdot}5\%$, $\sigma = 30$ b. The time of irradiation is very short compared to the half-life so that Equation 5.16 applies,

$$-\frac{dN_B}{dt} = N_A \sigma f \lambda_B t,$$

$$I_2 = \frac{6{\cdot}023 \times 10^{23}}{58{\cdot}5} \times \frac{75{\cdot}5}{100} \times 30 \times 10^{-24} \times 10^{10} \times \frac{0{\cdot}693 \times 10}{3 \times 10^5 \times 365 \times 24},$$

$$= 6{\cdot}15 \text{ dis s}^{-1} \text{ for } ^{36}\text{Cl}.$$

(*c*) ^{37}Cl. $t_{1/2}$ for ^{38}Cl $= 37{\cdot}3$ min. The abundance of ^{37}Cl $= 24{\cdot}5\%$, $\sigma = 0{\cdot}56$ b. The time of irradiation is long compared to the half-life and saturation will be produced so that the simple Equation (5.15) applies,

$$-\frac{dN_B}{dt} = N_A \sigma f,$$

$$I_3 = \frac{6{\cdot}023 \times 10^{23}}{58{\cdot}5} \times \frac{24{\cdot}5}{100} \times 0{\cdot}56 \times 10^{-24} \times 10^{10},$$

$$= 1{\cdot}41 \times 10^7 \text{ dis s}^{-1} \text{ for } ^{38}\text{Cl}.$$

The total disintegration rate of the specimen at the end of the irradiation would be $3{\cdot}39 \times 10^7$ dis s^{-1} and the activity of the ^{36}Cl can be ignored. The total activity would fall rapidly during the first few hours while the ^{38}Cl decayed, thereafter decaying with the 15·4 h half-life of ^{24}Na (see Experiment 24).

Case 4. A piece of gold foil weighing $8{\cdot}85 \times 10^{-3}$ g was irradiated in an unknown flux of slow neutrons for one hour. The activity of the gold foil was measured 46·5 h later and gave $5{\cdot}43 \times 10^4$ dis min^{-1}. What was the flux?

$t_{1/2}$ for ^{198}Au $= 2{\cdot}70$ d. The abundance of ^{197}Au $= 100\%$, $\sigma = 98$ b. First correct the disintegration rate of the gold for decay:

$$\ln I_0 = \ln I_1 + \lambda t$$

where I_0 and I_1 stand for the disintegration rate at the end of the irradiation and 46·5 h later respectively, and $t = 46{\cdot}5$ h. So, converting to common logarithms:

$$\log_{10} I_0 = 4{\cdot}7348 + \frac{0{\cdot}693 \times 46{\cdot}5}{2{\cdot}70 \times 24 \times 2{\cdot}303}$$

$$= 4{\cdot}9508,$$

and $$I_0 = 8{\cdot}93 \times 10^4 \text{ dis min}^{-1}$$

Next calculate the flux using Equations (5.16) as the time of irradiation is short compared to the half-life of ^{198}Au.
Rearranging (5.16):

$$f = I_0/N\sigma\lambda t$$

and:

$$N = \frac{6{\cdot}023 \times 10^{23} \times 8{\cdot}85 \times 10^{-3}}{197},\ \sigma = 98 \times 10^{-24}\ \text{cm}^2,$$

$$\lambda = \frac{0{\cdot}693}{2{\cdot}70 \times 24}\ \text{h}^{-1}$$

$$t = 1\ \text{h}$$

$$\text{so, } f = \frac{8{\cdot}93 \times 10^4}{60} \times \frac{197 \times 2{\cdot}70 \times 24}{6{\cdot}023 \times 10^{23} \times 8{\cdot}85 \times 10^{-3} \times 98 \times 10^{-24} \times 0{\cdot}693 \times 1}$$

$$= 5{\cdot}25 \times 10^7\ \text{neutrons cm}^{-2}\,\text{s}^{-1}.$$

Questions

5.1. Explain the meaning of exponential decay.

The disintegration of a radioactive nucleus is considered to be a matter of chance. How do you reconcile this idea with the observed regular law of decay?

In an experiment using an isotope of radon in a proportional gas counter the observed initial rate of decay was 90 pulses s^{-1}. After 60 s the rate had fallen to 45 pulses s^{-1}, and after 10 min to 4 pulses s^{-1}. The rate then very slowly decreased with a half-life of 10 h indicating that the gas counter was initially contaminated. From these figures calculate the half-life of the isotope.

5.2. Give a short account of some nuclear reactions induced by high energy particles, explaining why particles of high energy are required.

When deuterium is exposed to γ-radiation of energy 2·62 MeV the following reaction occurs:

$${}^2_1\text{D} + \gamma \rightarrow {}^1_1\text{H} + {}^1_0\text{n}$$

Measurements show that the proton and neutron produced in this reaction each have the same kinetic energy of 0·225 MeV. Calculate the mass of the neutron in a.m.u. (The isotopic masses are ${}^2_1$D = 2·0147, ${}^1_1$H = 1·0072 a.m.u. $c = 3 \times 10^8$ m s^{-1}, 1 eV = $1{\cdot}60 \times 10^{-19}$ J, 1 a.m.u. = $1{\cdot}660 \times 10^{-27}$ kg.) (Nottingham Physics Part 1.)

5.3. The following table gives the energies of the α-particles, and the corresponding half-lives, for 4 members of the 4n series.

Isotope	$t_{1/2}$	Energy (MeV)
^{228}Th	1·90y	5·42
^{224}Ra	3·64d	5·68
^{220}Rn	54·5s	6·28
^{212}Bi	60·5min	6·09

Use this information to obtain approximate values for the half-lives of the following isotopes: ^{232}Th (3·98 MeV), ^{216}Po (6·77 MeV), ^{212}Po (8·78 MeV).

5.4. Explain what is meant by the half-life of a radioactive isotope. Describe a method for measuring this quantity. Over what range of values of the half-life can your method be used?

A nucleus X with a half-life T_x decays to form the nucleus Y which is itself radioactive, having a half-life T_y. A sample initially only contains nuclei of type X. After a time t, the numbers of nuclei of type X and Y present are N_x and N_y respectively. Show that the activity (disintegration rate) of the Y nuclei reaches a maximum when $N_x/T_x = N_y/T_y$. Explain the observation that if T_y is very small compared with T_x, the activities of X and Y nuclei in the sample eventually become and then remain equal. (Oxford schol. Advanced Physics.)

5.5. One of the isotopes of gadolinium, ^{152}Gd, is radioactive. Its abundance in the naturally occurring element is 0·20%. A sample consisting of 5·0 g of metallic gadolinium was found, by approximate methods, to emit 114 α-particles every 4 h. Calculate the half-life of ^{152}Gd.

5.6. In the Harwell publication *Radioisotope Data* (1959 edition), the saturation activity due to ^{51}Cr, produced by neutron irradiation of ^{50}Cr, is given as 18 mCi g^{-1}. The activity after irradiation for four weeks is given as 8 mCi g^{-1}.

From these figures, estimate the half-life of ^{51}Cr.

5.7. For the process: $^{34}S + {}^{1}_{0}n = {}^{35}S + \gamma$, the saturation activity resulting from irradiation of a sulphur compound in one of the Harwell reactors is given as 550 μCi g^{-1}. The half-life of ^{35}S is 87·2 d.

Calculate the activity which would be produced in a sample irradiated (i) for 7 d, (ii) for 87·2 d. (Adapted from a Nuffield question.)

5.8. The abundance of ^{234}U in natural uranium is about 0·006%. Given that the half-life of ^{238}U is $4{\cdot}5 \times 10^9$ y, estimate the half-life of ^{234}U.

5.9. When $^{27}_{13}Al$ is irradiated with α-particles, the products from each nucleus are a neutron, and a nuclide which emits positrons to give the stable $^{30}_{14}Si$. Write 'nuclear' equations for these two processes.

The activity of the positron-emitter was found to change with time as follows. Plot log (count rate) against time, and determine its half-life.

t/min:	1	2	3	4	5	6	7	8	9	11	13
log (count rate)	5000	3841	2915	2220	1724	1293	977	745	617	361	209

(Adapted from a question in Physical Science A, 1968, of the Oxford Scholarship/Entrance Examination.)

5.10. The thermal fission of uranium in a nuclear reactor produces radioactive isotopes of the rare gases krypton and xenon. The γ-activity of a mixture of ^{88}Kr and ^{138}Xe was measured at various times. Another identical sample was passed over active charcoal to remove ^{138}Xe, and the activity of the ^{88}Kr alone was measured at various times. The results were as follows:

For $^{88}Kr/^{138}Xe$ mixture:

t/min:	0	10	20	30	40	50	75	100
Activity/counts s^{-1}:	200	161	135	115	100	94	80	68

For ^{88}Kr alone:

t/min:	30	60	90	120	160
Activity/counts s^{-1}:	88	78	69	61	52

(*a*) Plot log (activity) against time for the mixture, and comment on the result.

(*b*) Plot log (activity) against time for ^{88}Kr, and determine its half-life.

(*c*) From the graph in (*b*), determine the activity due to the ^{88}Kr at various times, and hence derive a set of figures for the activity of ^{138}Xe at various times.

(*d*) Plot log (activity) against time for ^{138}Xe, and determine its half-life.

(Adapted from a question in the O. and C. Chemistry A-level paper II, 1968.)

5.11. (*a*) Calculate the mass of that amount of metallic ^{238}U (half-life $4{\cdot}5 \times 10^{9}$ y) which has an activity of 1 Ci ($3{\cdot}7 \times 10^{10}$ dis s^{-1}).

(*b*) Calculate the activity, in curies, of 1 μg of metallic ^{216}Po (half-life 0·16 s).

6. Properties of the radiations and decay processes

6.1. *Properties of the radiations*

There are four types of radiation to be considered

(1) α-particles.
(2) β-particles.
(3) γ-rays.
(4) Neutrons.

6.1.1. *α-particles*, of mass 4 (atomic mass scale) and charge +2 are characterized by the intense ionization and consequently low penetration of the medium through which they are passing. The velocity of emission from the nucleus is approximately 10^9 cm s^{-1}, varying with the energy. The ionization produced increases as the α-particle slows down, rising to a maximum and then falling abruptly (Figure 6.1). To ionize a molecule of oxygen or nitrogen requires an energy transfer of about 32·5 eV. A 4 MeV α-particle (comparatively low energy) will therefore produce some 130 000 ion pairs in air before being stopped. This would correspond to about 3 cm of air at S.T.P. The equivalent thickness of aluminium would be approximately 0.01 mm. α-particles are characterized by a precise energy and therefore show a line spectrum. An α-active nuclide may emit several different groups of α-particles, each group having a definite energy (p. 95).

6.1.2. *β-particles*, which are high energy electrons or positrons, have a mass in either case of about 1/1800 of the mass of a proton. Both particles produce ionization of the medium through which they pass, but not nearly so intense per centimetre of path as α-particles. The penetration is considerable. A β-particle of moderate energy (0·7 MeV) will penetrate about 200 cm of air at STP, equivalent to about 1 mm of aluminium. The velocity of emission from the nucleus is very close to the velocity of light. When negative β-particles are slowed to thermal energies they are captured by a positive ion. Positrons, on the other hand, when reduced to thermal

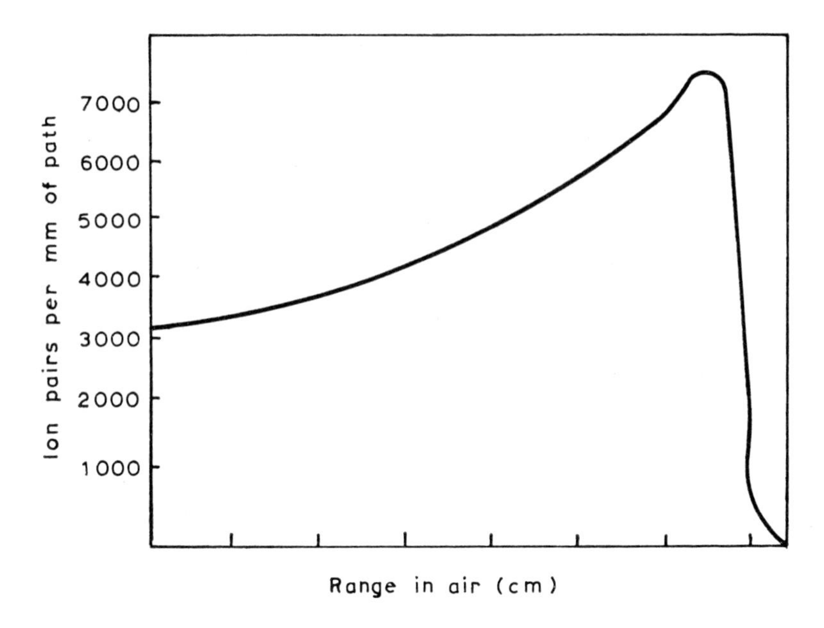

Figure 6.1. Ionization curve for an 8 MeV α-particle. The ionization increases as the α-particle slows down, rising to a maximum and then ceasing abruptly. This is characteristic of all charged particles and is well shown in photographic emulsions (see Figures E34(a) and b, pp. 240 and 241).

energies, collide with an electron with mutual annihilation and the emission of two 0·51 MeV γ-photons. γ-radiation of this particular energy is a characteristic of all nuclides which are positron emitters.

Unlike α-emitters, the energy spectrum of a β-emitter is continuous up to a maximum (Figure 6.2). After its discovery by Chadwick in 1914 this continuous β-spectrum posed a serious theoretical problem. That the atomic nucleus has discrete energy levels is shown by the line spectra of both α- and γ-emitters. The mean β-energy is about one-third of the maximum. These observations would imply that the law of conservation of energy did not apply to β-decay. There is a further difficulty. When an α-particle is emitted the α-particle and the nucleus recoil in exactly opposite directions as is required from the law of conservation of momentum. However, in the case of β-emission this is not so, and the tracks of the β-particle and the nucleus are inclined at an angle, the angle becoming greater with increasing β-energy until it is 180° at the maximum energy.

It seemed highly improbable that these conservation laws should be violated in the case of β-emission, and the Swiss physicist W.

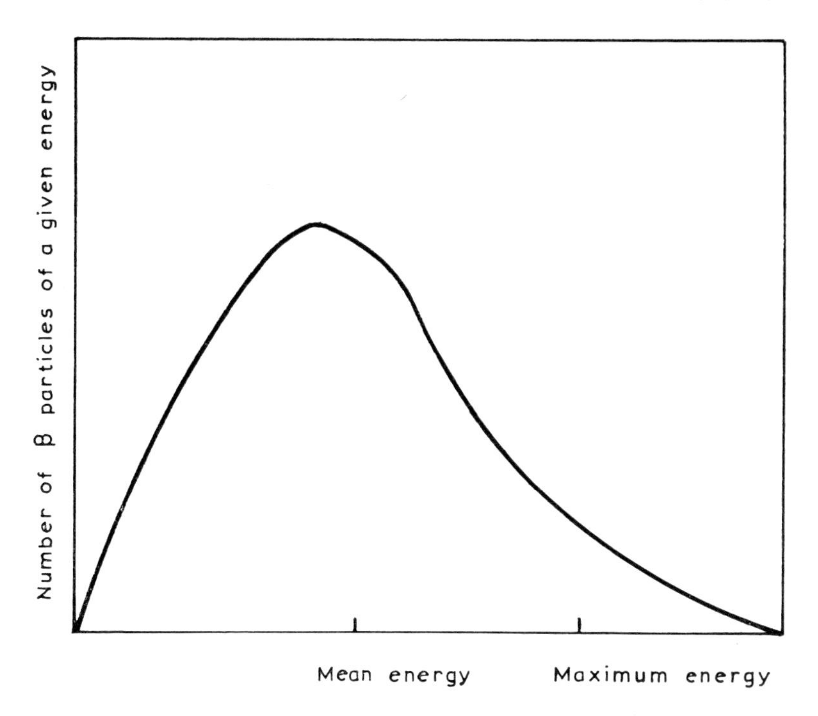

Figure 6.2. A typical β-spectrum. The spectrum is continuous to a maximum energy. The difference in energy between any given β-particle and the maximum is made up by the energy of the neutrino simultaneously emitted. The mean energy of all the β-particles is approximately one third of the maximum.

Pauli proposed that energy and momentum were conserved by the simultaneous emission of a neutral particle. This neutral particle was named by Fermi the neutrino, meaning 'little neutral one'. As it had not so far been detected it was necessary to assume that its rest mass must be very small or zero. An upper limit for the rest mass can be obtained by measuring the maximum energy of a β-particle and the rest masses of parent and daughter nuclide. From measurements involving the decay of ^{3}H the combined rest mass and kinetic energy of the neutrino emitted can be estimated as the equivalent of 200 eV. Since the rest mass of an electron is the equivalent of 0·5 MeV the rest mass of the neutrino cannot be more than 4×10^{-4} electron rest masses. It is now considered probable that the rest mass is zero. It has been shown that there are, in fact, two neutrinos, the anti-neutrino emitted in β^{-}-decay and the neutrino emitted in β^{+}-decay. Their only difference is in the direction of spin. (For the detection and further properties of both neutrinos the reader is referred to *Neutrinos* by G. M. Lewis in this series.)

Nuclides which are unstable due to too high a neutron/proton ratio decay by emission of a negative β-particle. Isotopes produced by bombardment with slow neutrons decay by this process except in the rare cases of neutron emitters. On the other hand, nuclides which have an excess of protons decay either by emission of a positron, or by capture of an electron, generally from the K shell, which is accompanied by the emission of a cascade of X-rays due to electrons moving inwards from outer shells to fill the gap.

6.1.3. *γ-rays* are electromagnetic radiation similar to X-rays, but of very short wavelength (high energy) and originating from the nucleus. When a nuclide emits an α- or β-particle the daughter nuclide produced is often in an excited state and rapidly loses its surplus energy by γ-emission. The energy is generally lost in a succession of very rapid steps by the emission of several γ-rays. This produces a γ-spectrum which is a line spectrum, i.e. discontinuous, unlike the β-spectrum. The energy is related to the wavelength by the Einstein equation $E = hc/\lambda$. Due to the absence of charge, γ-rays produce very little ionization (in air about 1·5 ion pairs cm^{-1} of path at STP) and consequently the penetration is very considerable. The absorption of γ-radiation by matter follows an exponential law. If the original intensity is a_0 and the intensity after passing through a thickness x cm is a, then

$$a = a_0 \exp(-\mu x)$$

where μ is the absorption coefficient. μ is a function of the density of the material. γ-rays of 1 MeV energy will be reduced to half their intensity by passing through 1 cm of lead. This would correspond to about 10 m of air. The absorption in air is so slight that the intensity of the radiation almost follows the inverse square law.

The absorption of γ-rays by matter is due to a variety of effects (Figure 6.3). The three principal effects are:

(*a*) *The photoelectric effect.* γ-rays, like all electromagnetic radiations, are emitted discontinuously in units of energy called 'quanta' or 'photons'. The photoelectric effect is due to the γ-photon giving up the whole of its energy to an electron, which is ejected from the atom to which it was attached with the energy of the γ-photon less the ionization energy. This effect predominates with γ-rays of low energy and in materials of high atomic number.

(*b*) *The Compton effect.* This is due to an elastic collision between a γ-photon and an electron. Energy and momentum are conserved, the electron recoiling at an angle to the incident photon which also recoils. Several collisions of this sort may occur until ultimately the photon is photoelectrically absorbed. The energy of the recoil electrons is continuous to a maximum which corresponds to a head-on collision, the photon being deflected through 180°. The Compton effect decreases with increasing energy.

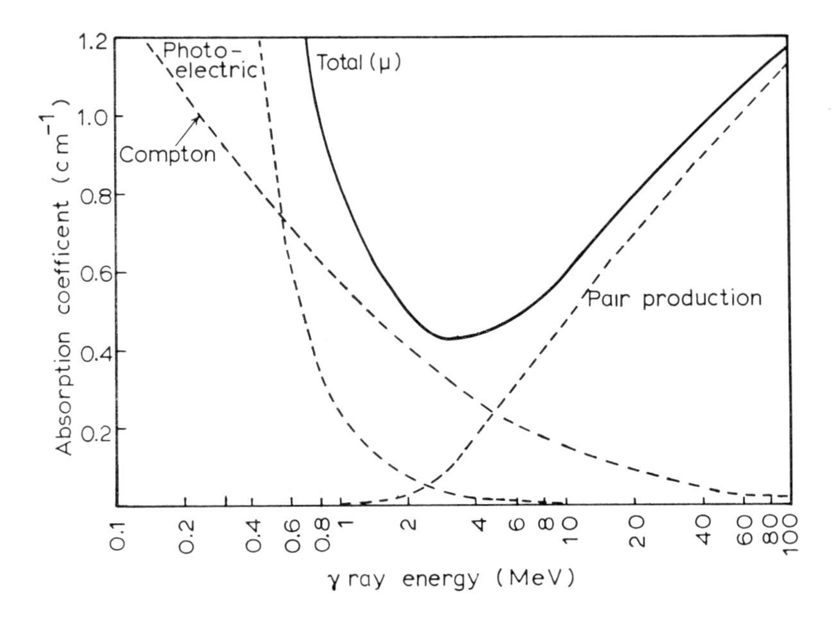

Figure 6.3. Total and partial absorption coefficients for γ-rays in lead. The Compton effect is the scattering of γ-rays with partial transfer of energy to electrons in the absorber. The photoelectric effect is produced by the transfer of the total energy of a γ-ray to an electron. Pair production, which predominates at high energies, is the conversion of a γ-ray into an electron–positron pair. The relative importance of each mode of absorption varies with the atomic number of the absorber. (From *Radioactive Isotopes* by W. J. Whitehouse and J. L. Putman, 1953, reproduced by permission of Oxford University Press.)

(*c*) *Pair production.* This is the creation of an electron-positron pair in the coulomb field close to an atomic nucleus. The photon disappears and its energy is distributed between the energy of the recoiling nucleus, the two electrons and the mass equivalent, which is $2m_0c^2$ where m_0 is the rest mass of each electron. This is equal to 1·02 MeV. For pair production this is the minimum energy for the incident photon. Pair production increases with increasing energy.

In addition to these three effects, nuclear reactions of the (γ, n) type may occur. In general these only happen with very high energy photons (6–8 MeV), but in one particular case ${}^{9}_{4}Be(\gamma, n)2{}^{4}_{2}He$, the energy required is only 1·63 MeV. This reaction is of practical application in the production of a laboratory neutron source (p. 141).

6.1.4. *Neutrons.* These particles have a mass of 1 and charge 0. They are unstable and decay ($t_{1/2} \simeq 10$ min) to a proton, an electron, and an anti-neutrino. This, however, very rarely occurs, as neutrons when reduced to thermal energies are easily captured by nuclei. Being uncharged, they produce no direct ionization in air and their penetration of matter is very considerable, particularly in materials of high atomic number. They are slowed down or 'moderated' by elastic collisions with nuclei; elements of low atomic number such as hydrogen, deuterium and carbon, are the most effective as moderators. Deuterium is much to be preferred to hydrogen as a moderator as hydrogen has a comparatively high capture cross-section (p. 24) for slow neutrons. Carbon in the form of graphite is much used as a moderator in nuclear reactors (Chapter 3). Neutrons vary in energy from several MeV down to thermal energies (0·05 eV). Fast neutrons constitute a considerable biological hazard owing to ionization produced in the body by collision with protons. Shielding commonly used consists of carbon or concrete. Thermal neutrons are less harmful biologically than fast neutrons (p. 159).

6.2. *Decay processes*

Nuclei have been observed to decay by the following spontaneous processes:

(1) α-emission.

(2) β^- or β^+emission or K-capture.

(3) γ-emission.

(4) Fission.

(5) Neutron emission,

A nucleus is energetically unstable to any given mode of decay provided its nuclear mass is greater than the sum of the masses of the decay products which would be formed. This is the factor which determines the type of decay. However, as in chemical reactions, the important factor is the rate of decay. This is determined by the energy which the particle to be emitted requires to penetrate the nuclear surface.

6.2.1. *α-decay.* All nuclei above $A = 140$ have negative binding energies with respect to α-decay. In fact this form of decay is very rare below $A = 212$. In classical mechanics an α-particle must overcome the coulomb barrier, but in terms of quantum mechanics there is a finite chance of an α-particle penetrating the barrier. This chance is called the 'penetrability'. The half-life is inversely proportional to the penetrability. In energy terms α-decay is normally from the ground state of the parent to the ground or an excited state of the daughter. This gives rise to the emission of groups of α-rays of different energies, the energy differences being small. The excited state of the daughter decays to the ground state with the emission of γ-rays. In certain cases nuclei formed by β-decay are so unstable to α-decay that α-particles are emitted before the nucleus has time to decay to the ground state by γ-emission. This accounts for the long range α-particles emitted by such nuclei. A good example is ^{212}Po formed by β-decay of ^{212}Bi. This gives out several different groups of α-particles, some of long range. ^{212}Po has $t_{1/2} = 10^{-7}$ s, showing its extreme instability (Figure 6.4).

6.2.2. *β-decay.* The nature of β-decay is complex. β-particles cannot exist in the nucleus but are formed by interchange of protons

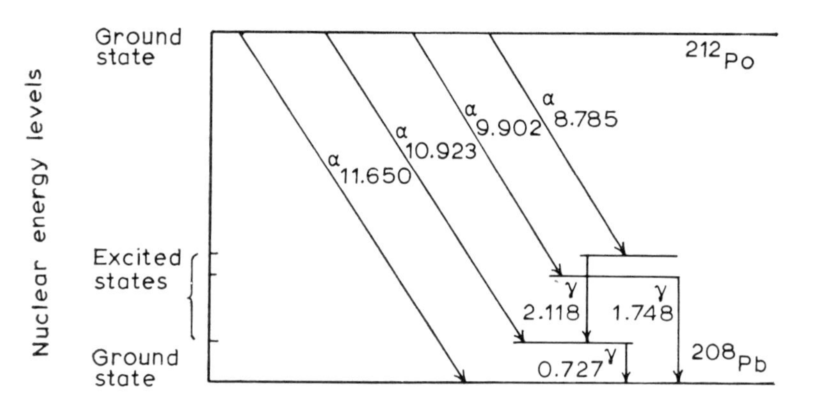

Figure 6.4. α-decay of ^{212}Po (the units are in MeV). This figure shows why α-emission is frequently accompanied by γ-emission since the product nucleus, in this case ^{208}Pb, may be in an excited state. The excess energy, equivalent to mass, must be emitted as a γ-ray (see also Question 6.1).

with neutrons or vice versa. The processes which occur in the nucleus may be represented as follows:

$$(a)\ n \rightarrow p^+ + e^- + \bar{\nu}$$

$$(b)\ p^+ \rightarrow n + e^+ + \nu$$

$$(c)\ p^+ + e^- \rightarrow n + \bar{\nu}$$

where $\bar{\nu}$ and ν represent an anti-neutrino and a neutrino respectively.

As mentioned previously, the β-spectrum is continuous and the neutrinos are included for energy balance. Mass and energy balance for these reactions may be written in the form of the following equations:

$$(a)\ m_Z \rightarrow m_{Z+1} + m_e + Q/c^2\ (\beta^-\text{-emission})$$

$$(b)\ m_Z \rightarrow m_{Z-1} + m_e + Q/c^2\ (\beta^+\text{-emission})$$

$$(c)\ m_Z + m_e \rightarrow m_{Z-1} + Q/c^2 + B/c^2\ (\text{k-capture})$$

which may be written:

$$m_Z \rightarrow m_{Z-1} - m_e + Q/c^2 + B/c^2$$

where m_Z represents the mass of the parent nucleus, $m_{Z\pm1}$ that of the daughter, m_e the mass of an electron, Q the combined kinetic energy of the β-particle, neutrino and any γ-decay energy, and B the binding energy of the captured electron in the nucleus.

Note that the mass of the daughter nucleus in process (b) must be at least 2 electron rest masses less than the mass of the daughter nucleus in process (c). Positron emission and electron capture compete with each other. Positron emission is much more frequent among the lighter elements, whereas K-capture is more frequent among the heavier elements.

6.2.3. *γ-decay*. This form of decay is restricted to nuclear isomers. In this case the nucleus is in an excited state only a few hundred keV above the ground state, and differs from the ground state by several units of angular momentum. Under these conditions γ-emission is highly inhibited, with the result that the decay rate of the excited state is measurable and often considerably slower than the decay rate of the ground state. Examples are ^{110m}Ag,† $t_{1/2}=280$d compared to ^{110}Ag, $t_{1/2}=24$s; ^{80m}Br, $t_{1/2}=4{\cdot}4$h compared to ^{80}Br, $t_{1/2}=17$ min; ^{104m}Rh, $t_{1/2}=4{\cdot}4$ min compared to ^{104}Rh, $t_{1/2}=43$s. ^{60m}Co, however decays with $t_{1/2}=10{\cdot}7$ min compared to ^{60}Co, $t_{1/2}=5{\cdot}3$ y. ^{60m}Co is exceptional in that it decays by

† The superscript m indicates an excited nuclear state.

β-emission, the β-ray being considerably more energetic (1·35 MeV) than that from ^{60}Co (0·31 MeV).

The γ-rays from nuclear isomers, being of fairly low energy, are generally internally converted, i.e. they eject an electron from the K-shell. This gives rise to X-rays as electrons from outer shells fill the gap. Thus nuclear isomers may be identified by the characteristic X-rays and very low energy β-particles (the ejected electrons from the K-shell).

6.2.4. *Fission.* All nuclei above $A=100$ have negative binding energies with regard to spontaneous fission (Figure 2.4). The rate of spontaneous fission is, however, unobservable below $A=232$, and even in this case ($^{232}_{90}Th$) the fission rate does not exceed 6×10^{-8} fissions $g^{-1} s^{-1}$. This would correspond to $t_{1/2}=10^{21}$ y, very much longer than the half-life for α-decay. For $^{238}_{92}U$ the half-life for fission is 10^{16} y, again much longer than for α-decay. For fission to occur a coulomb barrier has to be penetrated, similar to α-decay. In the case of ^{238}U the height of this barrier is 5 to 6 MeV. As in α-decay there is a finite probability that the barrier can be penetrated, but this probability is very small, as can be seen from the half-life. The barrier height decreases rapidly with increase in A. For this reason spontaneous fission as a mode of decay is, in effect, restricted to the very heavy synthetic elements.

(5) *Neutron emission.* This is a rare form of decay and occurs in the case of certain neutron-rich fission products (pp. 26, 27).

Question

6.1. The rest masses of $^{212}_{84}Po$, $^{208}_{82}Pb$ and $^{4}_{2}He$ are 211·98886, 207·97372 and 4·00264 respectively, in a.m.u. Show that the mass difference between the ^{212}Po and the combined masses of the ^{208}Pb and ^{4}He nuclei is approximately equivalent to the sum of the energies of the α-particle and γ-rays emitted (a.m.u. $=1{\cdot}6605\times10^{-27}$ kg, 1 MeV $=1{\cdot}6022\times10^{-13}$ J and $c=2{\cdot}997\times10^{8}$ m s^{-1}; see Figure 6.4.)

7. The detection and measurement of the radiations

7.1. *Introduction*

Two main methods are used today for detecting and measuring the radiations. The first depends on the ionization produced in matter by the passage of a charged particle, and the second depends on light emission (scintillations) produced by interaction of the radiation with a suitable phosphor. The second method is the most suitable for counting γ-photons since these only produce secondary electrons and relatively feeble ionization; this method is also used for counting very low energy β-particles such as those produced by ^{14}C, ^{35}S and ^{3}H. The first method is used for α-particles and moderate to high energy β-particles.

Ionization detectors will be discussed first. Ionization detectors can be divided into two types: those which simply count the particles and those which can measure the length of the track in the medium. The particle counters can be further subdivided into (i) those which measure the ionization produced in a gas, (ii) solid state detectors, and (iii) liquid scintillation counters.

7.2. *Gas Counters*

This counter consists in principle of an insulated central electrode in a suitable gas-filled chamber with a thin end window, a high potential being maintained between the central electrode and the outer wall (Figure 7.1). When an ionizing particle enters the chamber the ions produced will be attracted to the appropriate electrode and a minute current will flow, which can be observed with a suitable detector. In most cases the central electrode is made positive and electrons produced by the ionization of the gas move towards it and discharge.

The number of electrons produced by a given particle depends on the nature of the particle, the length of the track, and the applied voltage (Figure 7.2). It will be noticed that both the upper (α-particle) and the lower (β-particle) curve show a steep rise A B,

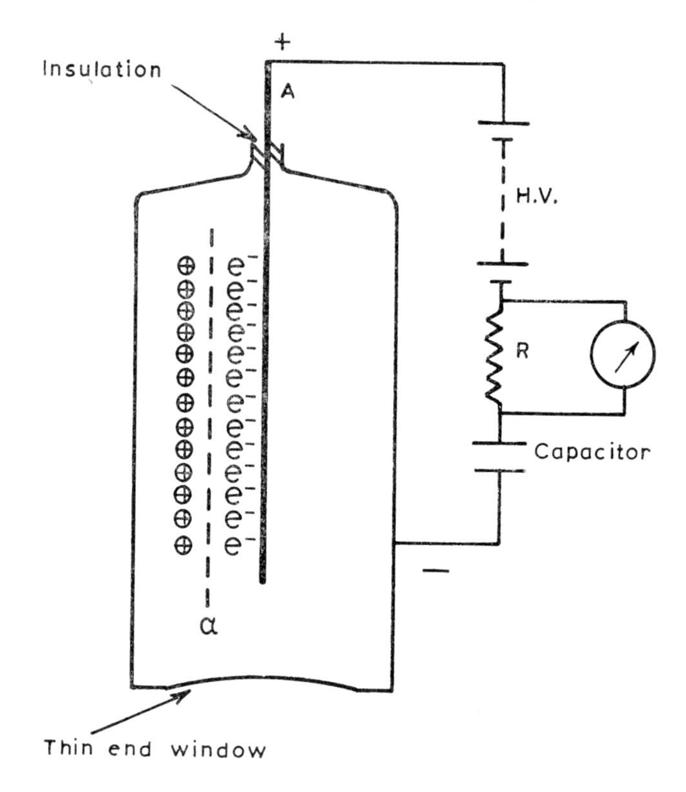

Figure 7.1. A gas counter. The charged particle α produces in its path a number of + ions and electrons. The electrons move rapidly to the central electrode, discharge and produce a momentary surge which causes a change of potential difference across the resistance R. The voltage 'swing' is recorded by the detector. The + ions move more slowly and have to be removed before another particle can be counted. In Geiger-Müller tubes a quenching agent, such as ethanol or bromine, is added, which reacts with the + ions and photoelectrons released from the wall of the tube.

A′ B′ at first, followed by a flat portion or 'plateau' B C, B′ C′. At low voltages many of the electrons produced in the primary ionization recombine before they reach the central anode, but as the voltage increases all the ions are collected. As the voltage is increased further the curves again rise steeply, C D, C′ D′. The primary electrons have now been accelerated to energies high enough to ionize gas molecules and produce secondary electrons. With still further increase in voltage more and more secondary electrons are produced. This process is called 'gas amplification', and it takes place within one millimetre of the anode, so that the

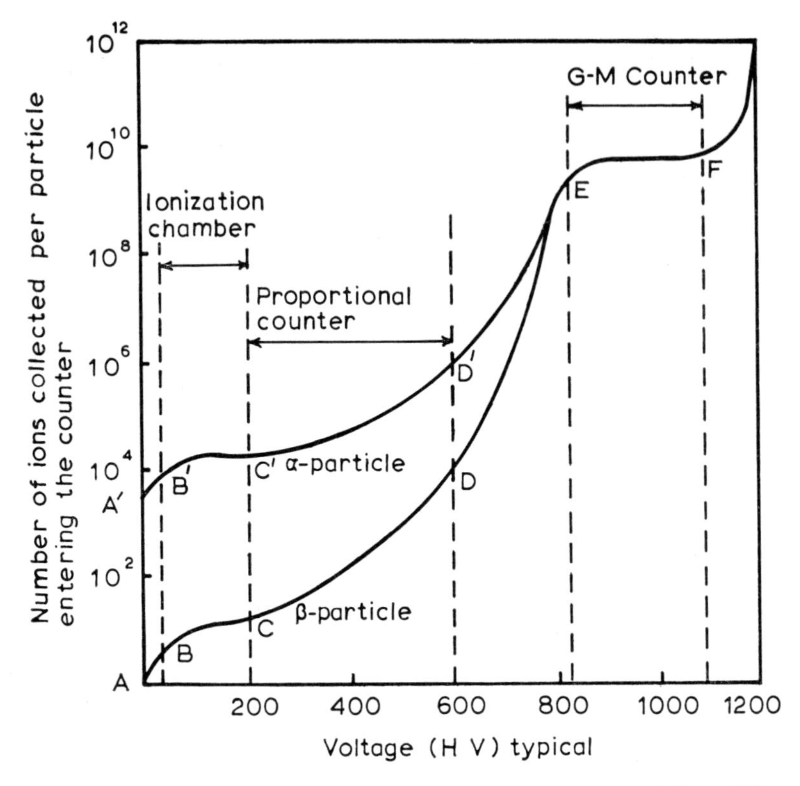

Figure 7.2. Ionization curves against applied voltage for α- and β-particles. At the points B,B′ all the electrons produced in the primary ionization are collected. Beyond C,C′ gas amplification occurs, but the number of secondary electrons produced is proportional to the number of primary electrons, the size of the pulse being determined by the nature of the particle entering the chamber. Beyond E, the Geiger region, saturation occurs and the size of the pulse is independent of the nature of the particle. EF is the Geiger plateau. (Reproduced from *Radioisotope Techniques* by R. T. Overman and H. M. Clark, copyright 1960, with permission of McGraw-Hill Book Co.)

position of the entering particle makes very little difference to the degree of amplification.

Between the points C D and C′ D′ the curves are a considerable distance apart, since the number of primary electrons produced by α- and β-particles differs widely. Over this region the number of secondary electrons produced is proportional to the initial number of primary electrons. Above D, D′ the curves run together and finally meet at E, producing another plateau E, F. At the voltages corresponding to this region the flow of electrons has become a veritable avalanche and the positive ions which move much more slowly are left as a screen around the anode, reducing the electric field intensity until eventually saturation occurs, so that no more electrons can be produced. This occurs at the point E, the start of the upper plateau. The upper plateau is known as the 'Geiger region'. Beyond F a state of continuous discharge occurs which may permanently damage the tube.

Figure 7.2 shows the conditions at which the three main types of ionization detector operate: the ionization chamber, the proportional counter, and the Geiger-Müller counter. Ionization chambers require high gain amplifiers, and their use will not be considered further here. As detectors of the integrating type they are used in the form of gold leaf electroscopes, portable β-γ-monitors, and gold plated quartz fibre electrometers. The pocket dosemeter which is widely used to measure radiation dosage (p. 159) is of the latter type.

Proportional counters are particularly valuable when it is necessary to distinguish different types of radiation entering the counter. The gas amplification factor may be of the order of 10^3, but the voltage pulse is still small enough to require the use of a high gain linear amplifier and also an extremely stable and reproducible high voltage power supply. Both these pieces of apparatus are very costly, and since proportional counters are not much used in radiochemistry they will not be discussed further. If the use of a proportional counter is required, the reader should refer to more advanced books on this subject, such as *Modern Radiochemical Practice* by Cook and Duncan.

The Geiger-Müller counter, in which the gas amplification factor is between 10^7 and 10^{10}, is the most commonly used instrument for both detection and counting. It is filled with a noble gas such as argon, generally at a pressure of one-fifth of an atmosphere, and also contains a 'quenching' agent such as ethanol or bromine which has a lower ionization potential than the gas filling. It is sensitive to α-, β- and γ-rays, though its efficiency for γ-rays is only about 1% of its efficiency for β-rays, since γ-rays produce very feeble ionization. When filled with BF_3 gas it can even be used for counting neutrons, though proportional counters lined with boron are generally employed. Owing to the relatively large voltage pulse produced a high gain amplifier is not required; a single stage of amplification is usually sufficient. As the 'plateau' is almost horizontal the high voltage power supply need not be so stable as for a proportional counter. This means that the cost of the electronic equipment which is required to operate the counter is much reduced.

G-M tubes, as they are called, are used in a variety of forms (figure 7.3). The end window type (a) can be used for α- and very soft β-particles, such as those produced by ^{14}C, if the window is made very thin (1·5–2 mg cm^{-2}). Such a counter is very fragile. A more robust type has a metal grid in front of the window which has a thickness of 4–5 mg cm^{-2}, and will count β-particles only. This counter is most useful in radiochemical work since the majority of radioactive isotopes are β-emitters. The cylindrical type (b) may be all metal, in which case it is only suitable for γ-emitters, or it may have a thin window as a strip throughout its length. In this form it is very convenient as a monitor for examining surfaces of benches for contamination. The liquid counter (c) is probably the most useful of all in radiochemical work, though it will not count α- or soft β-particles such as those produced by ^{14}C or ^{35}S, owing to the thickness of the glass envelope. Its capacity is generally 10 cm^3. The liquid flow counter (d) is useful in working with chromatographic or ion exchange columns to indicate the beginning and ending of the elution of a particular isotope. Many other types of G-M tube exist, but these are the ones most commonly used.

7.2.1. *Operation of the G-M tube.* When an ionizing particle enters the tube, a large number of electrons are produced, the discharge spreading along the entire length of the tube. The electrons relatively near to the anode are collected in a fraction of a μs, but the positive ions move outward much more slowly and it may take several hundred μs for them to reach the cathode. When they are relatively near the anode they form a space charge and until this space charge has moved away, outwards, the electric field strength will be too low to produce another pulse from an entering β-particle. The counter is thus rendered inoperative for a finite time, about 50–100 μs. This is known as the 'dead time'. This dead time varies with the applied potential and the age of the tube. Because of this, G-M tubes cannot be used for high counting rates. 5000–10 000 counts per minute is a suitable rate while 20 000–30 000 counts min^{-1} should be regarded as an outright maximum, and a considerable correction for lost counts would then be necessary (about 15%).

The electrons on striking the anode produce light photons which then strike the walls of the counter, releasing photoelectrons that can start a fresh avalanche. This would produce a state of continuous discharge. One function of the quenching agent is to absorb these photoelectrons. The positive ions would also produce photoelectrons on discharge at the cathode wall, and the second

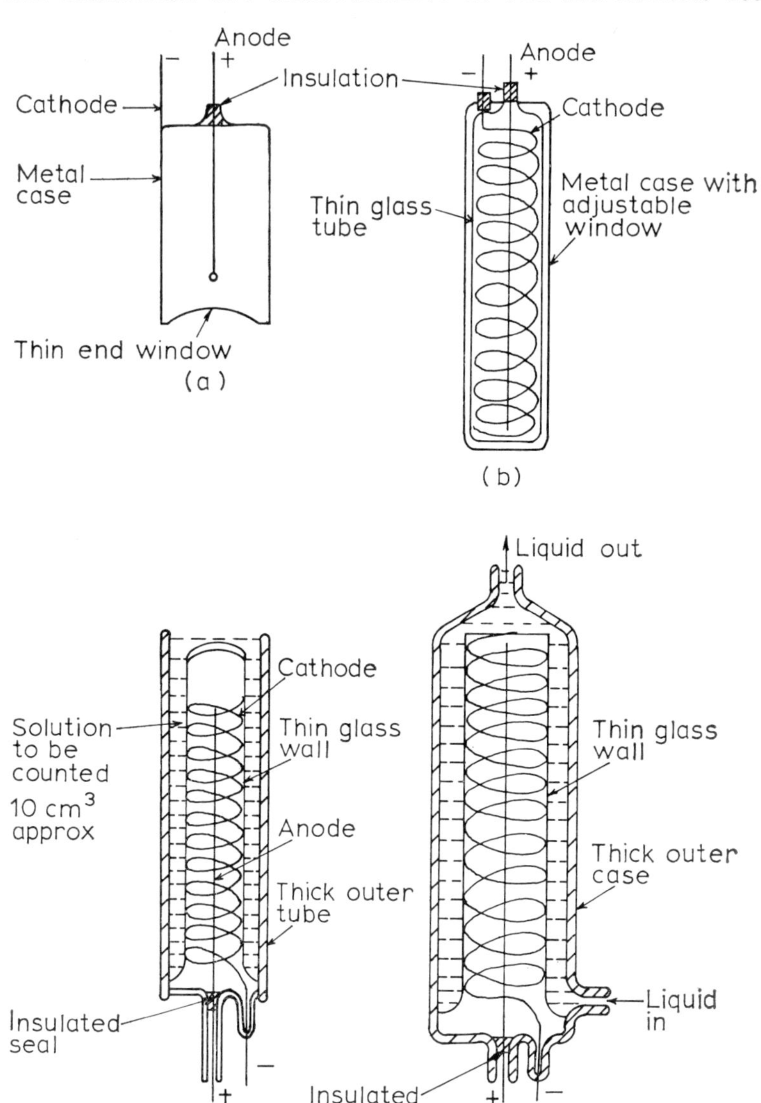

Figure 7.3. The main types of G–M tube. (*a*) is suitable for counting solid specimens of both α- and β-emitters, depending on the thickness of the window; (*b*) is useful as a bench monitor for β-particles above medium energy, or γ-rays; (*c*) is used for counting solutions of medium to high energy β-emitters; (*d*) is a flow counter similar in principle to (*c*) and is used for monitoring the effluent from chromatographic or ion exchange columns. The inner tubes are normally filled with argon containing a quenching agent, such as ethanol or bromine. Most tubes of types (*a*), (*b*) and (*c*) can be obtained with a plug fitting.

function of the quenching agent is to react with the positive ions. If ethanol is used as the quenching agent the ethanol ions formed by reaction with the positive ions and photo-electrons decompose on reaching the cathode and the decomposition products have insufficient energy left to start another discharge. The life of the counter is limited due to decomposition of the quenching agent. Counters containing ethanol (organic quenched) have a life of about 10^8 counts, and require an operating voltage of between 1000 and 1600 V. With bromine as the quenching agent the life of the counter is, in theory, infinite, since the bromine ions form free bromine atoms on discharge which recombine to form bromine molecules. Halogen quenched counters have this great advantage over organic quenched counters, and in addition work at an operating voltage of 400 to 600 V. The plateau of the halogen quenched counter is slightly shorter and the efficiency is slightly lower than the organic quenched counter, but in addition to requiring a much lower operating voltage, halogen quenched counters, unlike organic quenched counters, are not permanently damaged by applying, momentarily, an excessively high voltage. For student use they are to be preferred.

7.2.2. *Background.* All particle counters have what is called a 'background', i.e. they show a count when no radioactive substance is in close proximity to the counter. This is due to natural radiation from building materials which contain traces of thorium, from the air which contains small amounts of radon and its decay products, and from cosmic rays. The background can be considerably reduced by surrounding the counter with a lead shield, commonly called a castle. Most end window G-M tubes have a background of about 12 counts min^{-1} when shielded with 2·5 cm of lead. Liquid counting tubes, owing to their greater length, have a higher shielded background, generally of the order of 20 counts min^{-1}.

7.3. *Solid State Detectors*

These are essentially ionization chambers filled with a solid instead of a gas. This type of counter has certain advantages over the gas-filled counters in that high energy particles can be stopped completely within a comparatively small volume. The solid used is a semi-conductor, generally silicon or germanium. A charged particle entering the semi-conductor raises the energy of electrons into the conduction band and the electrons travel rapidly to the positive electrode. The positive 'holes' left by the electrons appear to travel to the negative electrode, though in fact the apparent movement is caused by electrons moving in to fill the hole from the neighbouring atoms. The only actual movement is that of electrons. For this reason the mobility of the positive holes is high, about one-third that of the electrons—a very much higher mobility than the positive ions in gas counters.

From the nature of the material a semiconductor must pass a current when an electric field is applied across it. This current is of the order of 0·1 A for a 1 cm cube of high purity silicon with an applied potential of 1000 V. This 'leakage' current can be reduced to a very small value by use of a 'p–n' junction. A p-type semiconductor has an excess of positive holes which are produced by the introduction of traces of elements such as boron or gallium, which are electron acceptors. An n-type semi-conductor is the opposite, i.e. it has an excess of electron carriers which are produced by introducing traces of elements such as phosphorus, arsenic or antimony, which are electron donors. A thin layer (10^{-3} mm) of p-type silicon is arranged on a piece of n-type silicon, with an evaporated gold contact on the p-type surface and evaporated aluminium contact on the n-type (Figure 7.4). A reverse bias is applied—on the p-type, and on the n-type. The positive holes are attracted towards the negative side and the electrons towards

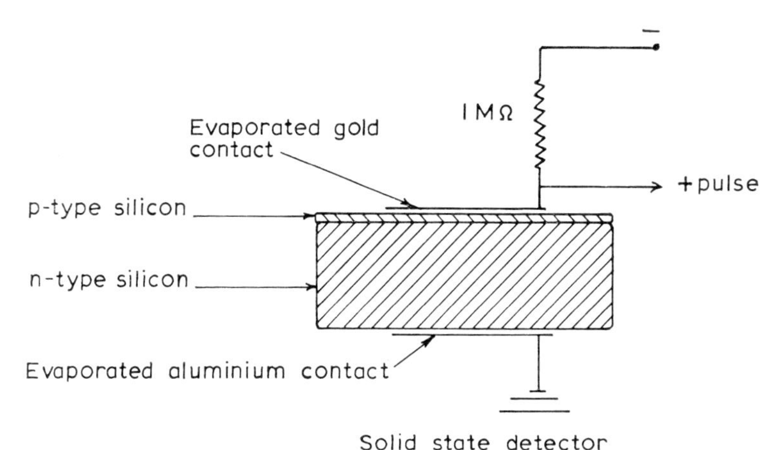

Figure 7.4. Solid-state detector (p–n junction). The p-type consists of a thin layer of silicon (or germanium), containing traces of electron acceptors (boron or gallium) and has an excess of positive 'holes'. The n-type, which forms the bulk of the detector, contains traces of electron donors (phosphorus, arsenic or antimony), and has an excess of electron carriers. The reverse bias produces a depletion layer of high resistance which becomes conducting when a charged particle enters it.

the positive side, with the result that a ‘ depletion ’ layer of high resistance is produced between.

When a charged particle enters the depletion layer, it immediately produces free electrons and positive holes, which are swept towards the respective electrodes and give rise to an electrical pulse. The magnitude of this pulse depends on the number of electrons and positive holes produced, and this depends on the energy of the particle. Solid state detectors are therefore useful as spectrometers.

The depth of the depletion layer depends on the resistance of the material and the applied bias voltage. With a resistance of 1000 ohms per cm depth and a reverse bias of 10 V the depletion depth would be about 50 μm, which would completely stop an 8 MeV α-particle, but only an 80 keV electron, i.e. a very low energy β-particle. With resistances of 10 000 $\Omega\ cm^{-1}$ and a bias voltage of 500 V a depletion layer 1·1 mm thick would be produced. This would stop β-particles of 0·6 MeV. It should be clear that to produce solid state β-spectrometers very high purity silicon is required and high bias voltages. The magnitude of the output pulse is of the order of 1 mV, and high gain amplifiers are required.

If the device is to be used simply as a particle detector it is not necessary for the depletion depth to be equal to the range of the incident radiation. The radiation, however, must deposit sufficient energy to produce a signal in excess of the noise level of the detector. A discriminator in the subsequent electronics can be set to cut out this noise level. The counting of α-particles, protons and other comparatively massive particles is relatively simple, but to count β-particles it is necessary to have a depletion depth of about 0·2 mm. This can be achieved with medium purity silicon of 3 $k\Omega\ cm^{-1}$ resistance, and with a reverse bias of about 50 V; the efficiency is low.

Solid state detectors can be used as thermal neutron detectors by fitting a foil of boron or lithium over the gold electrode. The neutrons liberate α-particles from the foil; these are counted by the detector. For fast neutrons, ‘ knock-on ’ protons are detected by having a cap of hydrogenous material over the detector.

A recent development is the lithium-drifted detector. Lithium is an n-type donor and lithium atoms may be passed (drifted) into p-type silicon at 150 °C under reverse bias. The mobility of the Li^+ ions is quite large and with an applied p.d. they move towards the negative electrode. The action is continued until the current drops almost to zero, which means that the lithium is distributed to compensate for the positive holes. By this means depletion layers of considerable depth may be produced which will stop quite high energy β-particles, and can therefore be used for β-spectrometry. They can detect γ-rays from the Compton scattered electrons. Owing to the relatively low charge on the silicon nucleus the probability of photoelectric interactions is small.

Solid state detectors have a number of advantages over gas counters. They are particularly valuable for counting positively charged high energy particles, and for spectrometry of such particles. Their recovery time is much shorter than a G-M tube, so that they are suitable for high counting rates. When used as detectors, they operate at a much lower voltage than a G-M tube and they have a very low background. Their efficiency as γ-counters is probably as good as a gas counter, but cannot be compared with a scintillation counter. They are useful as monitors for α- and β-emitters and for neutrons, and their small size and comparative robustness is an added advantage. They are more expensive than G-M tubes and require more refined electronics, and spectrometers require refrigeration with liquid nitrogen.

7.4. *Scintillation Counters*

The scintillation counter in the form of a zinc sulphide screen was one of the earliest types of detector, and was much used by Rutherford in his classical researches (p. 10). In its original form it was limited to very slow counting rates and was only suitable for α-particles and protons, since the scintillations produced by β-particles and γ-photons are much too feeble to be detected with the naked eye. The modern scintillation counter consists of a suitable phosphor used in conjunction with a photomultiplier tube. It is the invention of the photomultiplier tube which has made possible the development and wide application of scintillation counting.

7.4.1. *Photomultiplier tube* (Figure 7.5). This consists of a light sensitive photocathode, i.e. photocell, which is enclosed in an evacuated tube containing a series of electrodes (dynodes). A potential of the order of 100 V positive is maintained between the photocathode and the first dynode, and between each succeeding dynode. The photocathode is generally made from a caesium–antimony alloy and the dynodes of a beryllium–copper alloy; the dynodes are generally slatted like a venetian blind. The highest sensitivity of the photocathode is in the blue region of the spectrum. When a photon is emitted by the phosphor, this photon may eject an electron from the photocathode. The efficiency of this part of the process is between 5 and 10%. The ejected photoelectron is accelerated towards the first dynode where it releases about three secondary electrons, the number depending on the applied voltage. These in their turn are accelerated to the second dynode, where the process is repeated. Most photomultiplier tubes have eleven dynodes, so that the overall amplification is 3^{11}, i.e. several million electrons, from each photon detected, are discharged on to the anode. This causes a fast rising voltage pulse to appear at the anode capacitor (Figure 7.5), which is fed to a suitable amplifier and from there to a counting system. Some thermionic emission occurs in the photomultiplier tube producing small voltage pulses, and this is commonly called noise. To reduce or eliminate the noise it is necessary to use a discriminator between the photomultiplier and the amplifier so that pulses below a certain level are rejected. Owing to the extreme sensitivity of the photomultiplier it must be enclosed in a light-tight tube. Any stray light reaching the photocathode when the high voltage is applied would destroy the photomultiplier tube.

7.4.2. *Phosphors.* The type of phosphor to use depends on the radiation it is proposed to detect. For α-particles zinc sulphide activated with a trace of silver is the most suitable. The zinc sulphide in powder form is fixed as a thin layer on to a sheet of Perspex and covered with a very thin sheet of aluminium foil to exclude light. The clean side of the Perspex is placed in contact with the photomultiplier tube. Anthracene in the form of a thin crystal can be used for counting β-particles, but liquid phosphors are much more efficient (p. 111).

One of the most important uses of the scintillation counter is for counting γ-photons. The most suitable phosphor is the sodium iodide crystal activated with a trace of thallium, though organic phosphors can be used and are much cheaper. The advantage of sodium iodide is high efficiency, particularly at low γ-energies, and that the light emission is proportional to the energy of the γ-photon, so that it can be used for γ-spectrometry (p. 132), whereas organic phosphors cannot because they only detect through Compton scattering. Clear transparent sodium iodide crystals of large size can be grown and the larger the size the greater the efficiency, particularly for γ-spectrometry. Sodium iodide is hygroscopic, so it must be encased in a cylinder of aluminium with a clear plastic window which is in intimate contact with the face of the photomultiplier tube. Good optical contact is essential, and a trace of silicone oil is generally smeared on the two surfaces. The background of a scintillation counter is much higher than that of a G-M tube owing to the sensitivity of the crystal. The background noise can be reduced by cooling.

If the number of counts obtained with a steady source is plotted against the applied voltage the curve increases steeply at first and

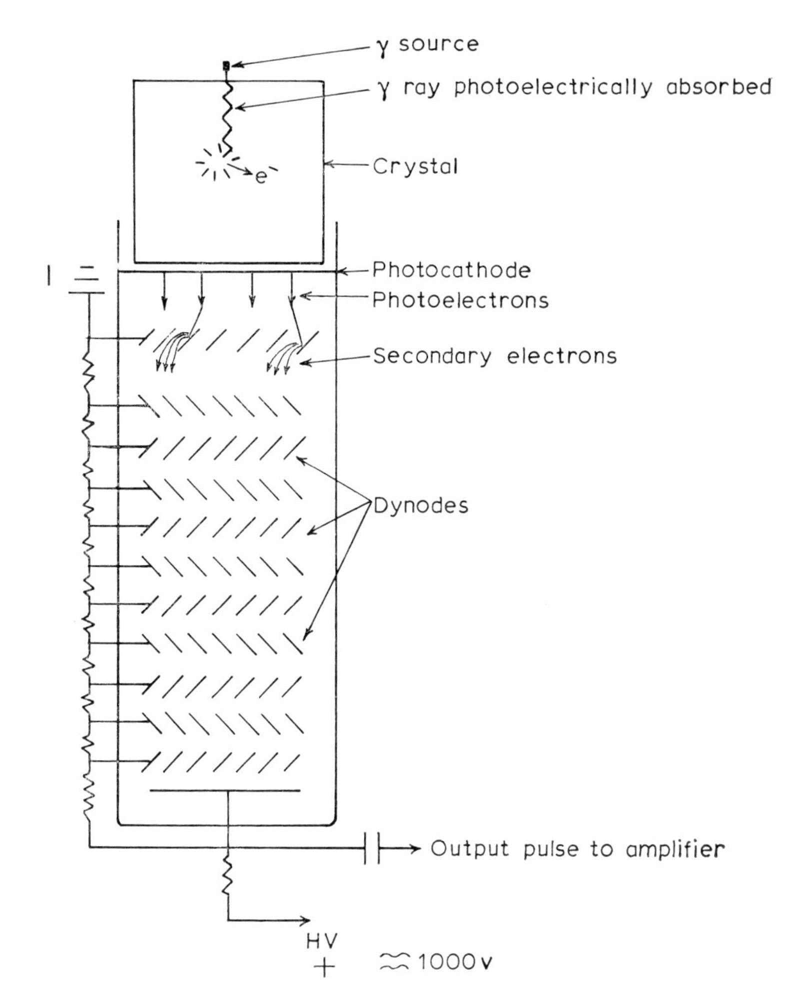

Figure 7.5. Photomultiplier tube and crystal. This is the normal arrangement for high sensitivity γ-counting. Low energy β-particles may also be counted if the crystal is replaced by a phial of a liquid scintillator. The whole must be enclosed in a light tight tube. The γ-ray is shown being photoelectrically absorbed. (For other types of interaction see Figure 9.5.) The light photon strikes the photocathode and the photoelectrons produced are accelerated to the first dynode, where each produces typically three more electrons. With 11 dynodes an amplification of 3^{11} is commonly obtained. The pulse is fed to an amplifier and from thence to a counter.

then more slowly, in some cases showing a short horizontal section (Figure 7.6) like the plateau of the characteristic curve for the G-M tube. In most cases the plateau slopes fairly steeply and for this reason it is necessary to use a very stable HV. Before using the scintillation counter, it is advisable to plot counts against applied voltage for the background, and then a similar curve for source plus background. By comparing these two curves the optimum condition for counting can be obtained, i.e. the largest value for R_s/R_b, where R_s is the count rate for the source and R_b the count rate for the background. It is advisable to plot these curves for different discriminator voltages and different amplifier settings.

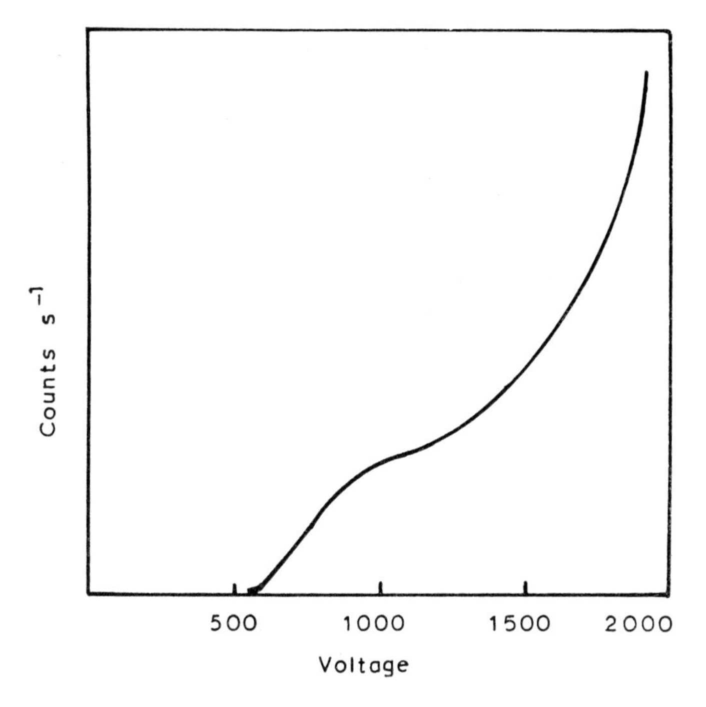

Figure 7.6. Typical γ-scintillation counting curve. The plateau is generally much shorter and slopes more steeply than the Geiger plateau and for this reason the high voltage supply must be much more stable than for Geiger counting. It is desirable to work about the middle of the plateau if possible.

Beyond the plateau the slope of the curve again increases sharply, and damage to the photomultiplier tube will occur if the safe working voltage is exceeded. Normally this is about 2000 V, but it is rarely necessary to use more than 1600 V (see Experiment 9).

7.4.3. *Liquid scintillation counters.* These instruments are a comparatively recent development. They consist of a photomultiplier tube with a glass or silica phial in optical contact with the upper end. This phial contains a liquid scintillator such as 2,5-diphenyloxazole (POP) in methylbenzene. A small quantity of the substance to be counted is dissolved in the liquid scintillator. The whole apparatus is completely enclosed to protect it from light.

Liquid scintillation counters are principally used for counting low energy β-emitters such as ^{14}C, ^{35}S and ^{3}H. The advantages over G-M tubes are considerable and very high counting efficiencies are obtained, of the order of 90% for ^{14}C. (^{3}H cannot be counted with a G-M tube. The only alternative to a liquid scintillation counter is a gas flow proportional counter.) To obtain the best results it is necessary to reduce the background to a minimum so that high voltages can be used with a low discriminator setting. High quality photomultiplier tubes are required and the tubes must be cooled. For ^{14}C and ^{35}S, cold water can be circulated through a jacket, but for ^{3}H counting the best results are obtained with a refrigerant.

As the β-spectrum is continuous the low energy end will be cut out by the discriminator, and the higher the discriminator setting the lower will be the counting efficiency. The shape of the β-spectrum curve depends on the voltage applied to the tube. The higher the voltage the greater the counting efficiency provided the amplifier does not cut off the upper end of the spectrum; high voltages, however, increase the background. To obtain optimum counting conditions it is necessary to plot counts against applied voltage for both source and background, as for γ-counting, but for different amplifier gain and discriminator settings. The background count must be made with a phial containing the pure liquid scintillator in position. For counting weak specimens the counting conditions should be adjusted so that R_s^2/R_b is a maximum, where R_s stands for the count rate due to the sample and R_b that due to the background. Typical spectrum curves are shown in Figure 7.7.

Liquid scintillation counters require a very stable high voltage (HV) and a high gain, non-overloading amplifier, considerably higher than for a γ-counter. A non-overloading amplifier is necessary because cosmic rays entering the scintillator will produce such large pulses that a normal amplifier could be blocked for several hundred μs with consequent loss of sample counts. Amplifier settings of 500 to 1000 are commonly used. It is also necessary to have a device for inserting and removing the phials without allowing any light to reach the photomultiplier. Alternatively, the HV must be switched off before the phial is inserted or removed. There must also be an arrangement to supply a trace of silicone oil to ensure optical contact between the base of the phial and the top surface of the photomultiplier. The phials are normally made of silica as most glasses are slightly phosphorescent. The base of the phial must be optically worked to make a good contact, and the outer side of the phial should be aluminized to act as a reflector.

There are certain problems specifically associated with liquid scintillation counting. The first concerns dispersion. It is essential to maintain uniform dispersal of the substance to be counted

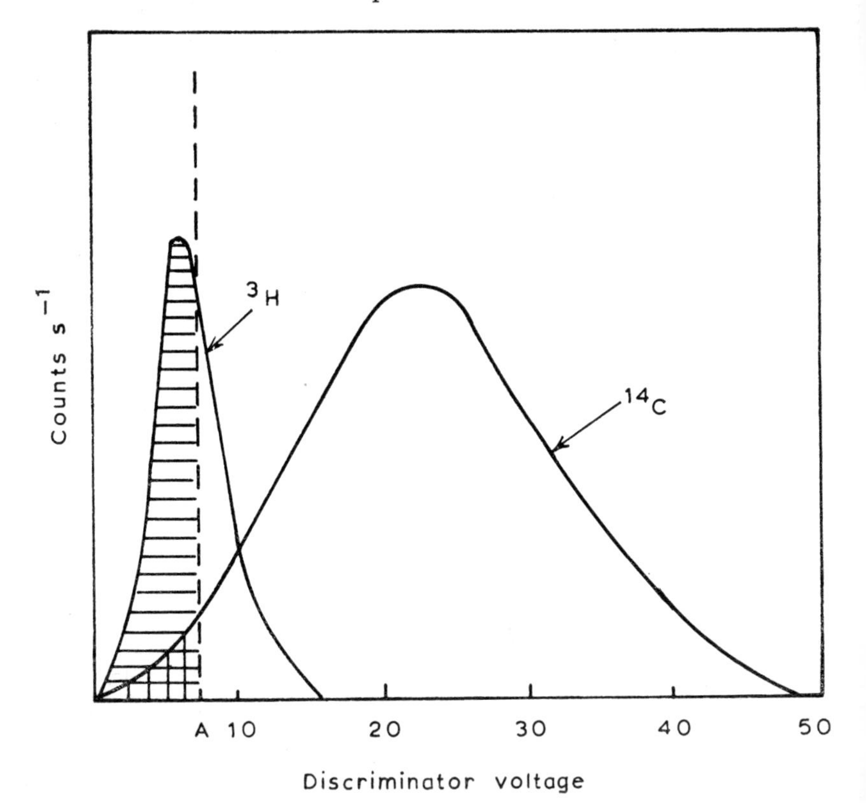

Figure 7.7. β-spectra as observed with a liquid scintillation counter. These curves are obtained by using an upper and lower discriminator, set to a difference of 1 or 2 V. The ' window ' is moved across the spectrum, making a series of counts against the discriminator voltage. To obtain maximum efficiency the lower discriminator is fixed at a point A, to minimize the background, and the upper discriminator is fully opened. With adequate cooling of the photomultiplier tube, efficiencies of over 90% may be obtained for ^{14}C, or 40% for ^{3}H. The ' lost ' counts are shown by the shaded part of the curve (see Experiment 12 (2)).

throughout the liquid. Organic compounds are generally soluble in methylbenzene, but with aqueous solutions it is necessary to use a dioxan solution of the scintillator. Dioxan will take up to 10% or more of water by volume without causing precipitation. If precipitation does occur the efficiency will fall considerably.

The second is phosphorescence of the scintillator. Unless the operations are conducted in subdued light the phial containing the scintillator must be dark-adapted by keeping it in the dark before counting. The time required varies from 5 to 20 min according to the type of scintillator used.

The third problem and the most serious one is quenching. Quenching is a process which interferes with the production of light in the liquid scintillant, and its transmission to the photomultiplier tube. There are two types of quenching; colour quenching and chemical quenching. Colour quenching is due to absorption of some of the light by coloured substances, and results in a lowering of the energy of the radiation transmitted to the photomultiplier tube. Chemical quenching is caused by interference by dissolved substances with the transfer of energy from the emitted radiation to the phosphor, and consequent degrading of the energy by processes which do not produce emission of light. Dissolved oxygen produces chemical quenching and must be removed by bubbling pure nitrogen through the liquid scintillant before use; liquid scintillators should always be stored under nitrogen. The degree of quenching is roughly proportional to the concentration of the quenching agent, but in general it is variable and very difficult to predict. Quenching causes a fall in efficiency by shifting the observed β-spectrum in the direction of lower energy (Figure 7.8). Corrections must therefore be applied due to the decreased efficiency of the counter. It is wise to assume that some quenching always occurs and act accordingly.

The effect of quenching can be studied by plotting a β-spectrum of carbon-14 using a pulse height analyser, and scanning the spectrum with a 1 or 2 V window. The carbon-14 should be in a chemical form which does not cause quenching. Labelled hexadecane, supplied by the Radiochemical Centre, Amersham, is very suitable for this purpose. The high voltage and amplifier gain are adjusted to suitable values and the β-spectrum plotted, first in the absence of a quenching agent, and then after adding successive quantities of tetrachloromethane (carbon tetrachloride) which is a powerful quenching agent. Curves similar to those shown in Figure 7.8 will be obtained (Experiment 12).

There are four main methods used to correct for quenching: external standard, internal standard, quench curve and channels ratio.

External standard. This consists in using an external γ-source and making two counts on each sample with and without the external source in position. The count produced by the γ-source with a phial of pure scintillant in position must be known, as also must the background count. If the sum of the counts produced by the sample and the external standard independently is greater than the count produced by the sample and external standard together, quenching has occurred. The principle of correction assumes that the degree of quenching of the external standard is the same as for the active substance. The method is not very accurate, due to uncertainty in positioning the external standard and possible variation in the volume of the scintillant.

Internal standard. Again, two counts are needed, but in this case, after taking the first count with the sample alone, the phial

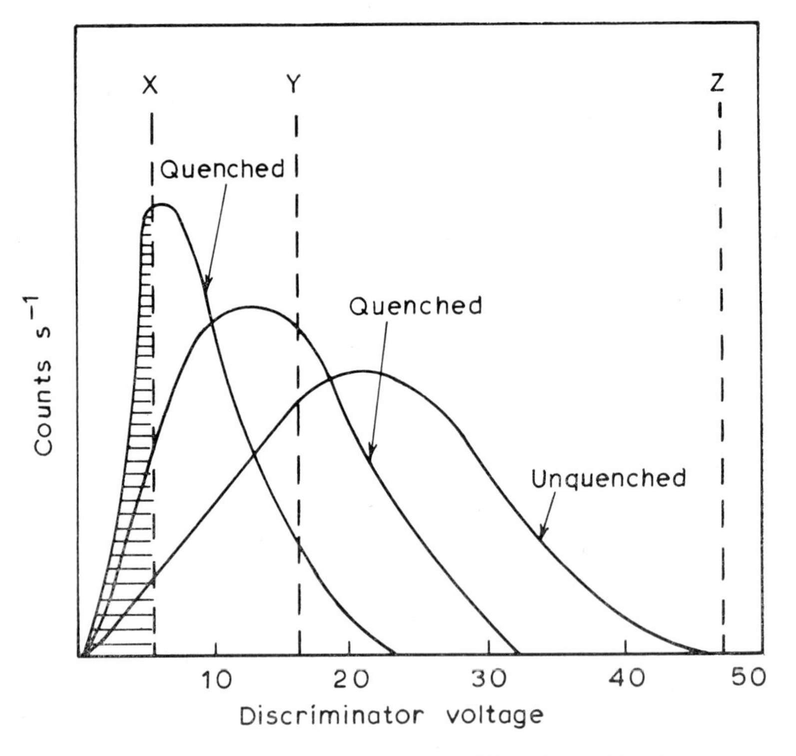

Figure 7.8. Quenching—typical curves for ^{14}C. Quenching is a process which interferes with the production of light in the liquid scintillator, and its transmission to the photomultiplier tube. It may be caused by the presence of coloured substances, or by substances which interfere with the transfer of energy to the phosphor. The effect is to reduce the efficiency of counting by causing a shift of the β-spectrum to the left.

must be removed from the castle, and a known quantity of the same isotope as the one being counted must be added, but in a chemical form which is known not to cause quenching. The principle of correction is exactly the same as for the external standard. The method is accurate, particularly for heavily quenched samples, but errors can be caused in measuring the small amount of internal standard needed. It is also tedious and time consuming.

Quench curve. In this case successive aliquots of the active substance are added to a given volume of scintillator in a phial, and a graph plotted of counts per minute against volume of substance added. If the graph is linear there is no quenching. If the counts per minute do not increase linearly a correction can be determined for any given volume of the active substance, provided the first part of the graph is linear (Figure 7.9 (*a*)). A better alternative is to add some of the isotope to a phial of scintillator in a chemical form which does not cause quenching and then add successive aliquots of an unlabelled form of the substance to be counted (Figure 7.9 (*b*)).

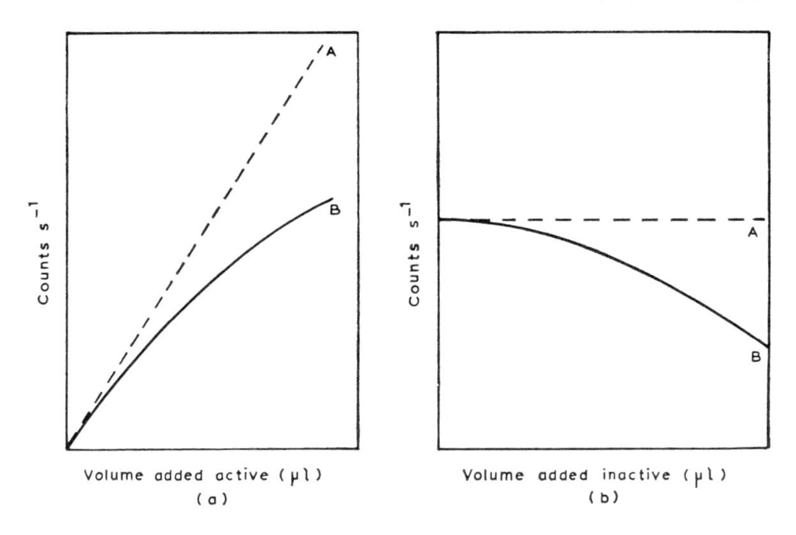

Figure 7.9. Quench correction curves. (*a*) Shows one method of correcting for quenching. Counts s^{-1} are plotted against the volume of a solution of the labelled substance to be counted. The volume is quoted in microlitres (μl) as the small syringes supplied for this type of work are graduated in these units. In the absence of quenching the graph is a straight line. The correction for any given volume is determined by the difference between curves A and B. (*b*) Shows an alternative method. Some of the isotope to be counted is added in a chemical form which does not cause quenching and a count made. Successive aliquots of an unlabelled form of a solution of the substance to be counted are then added, further counts being made. The correction is determined as before.

This method has the advantage that once a quench curve has been drawn only one measurement need be made on each specimen. The volume of the specimen must be known.

Channels ratio. The principle of this method depends on the fact that quenching causes a shift of the whole β-spectrum towards lower energies (Figure 7.8). If different sections of the spectrum are counted the ratio of the two counts will vary according to the degree of quenching. This process can be carried out using an upper and a lower discriminator. If two scalers fitted with discriminators are available both counts can be made simultaneously. With a single instrument it is necessary to make two counts on each specimen.

A known amount of a standard preparation of the isotope to be used must be added to the scintillator, the quantity added being such that it will not cause quenching. The HV and amplifier gain are adjusted so that the upper end of the spectrum does not exceed the range of the instrument. A count is then made over the full range with the lower discriminator set at X and a second count, simultaneously if two instruments are available, with the upper discriminator at Y (Figure 7.8). Let these counts be XZ and YZ respectively. From the known activity of the specimen and the count XZ the efficiency is determined. The ratio YZ/XZ is called the channels ratio. Successive quantities of a quenching agent such as tetrachloromethane are added, the efficiency determined, and the channels ratio YZ/XZ observed. The efficiency is then plotted against the channels ratio (Figure 7.10).

Once this curve has been determined it can be used to measure the degree of quenching, and hence the efficiency of any preparation of the isotope which is being used, irrespective of the quantity added. There is one further advantage of this method. A slight fall in the high voltage causes the same effect as quenching. Thus a change in efficiency due to slight variation in the high voltage supply can also be determined by this method. The channels ratio is probably the most useful method of all (Experiment 12).

The last problem is chemiluminescence due to chemical interaction between the substance and the phosphor, producing light photons. Unlike the scintillations produced by β-particles which consist of several simultaneously produced photons, chemiluminescence is a single photon process. To correct for this it is necessary to use a two channel instrument which consists of two photomultipliers, two amplifiers, two high voltage supplies and a coincidence circuit.

In addition to the counting of low energy β-particles, medium energy β-particles, such as those emitted by ^{131}I, can be counted

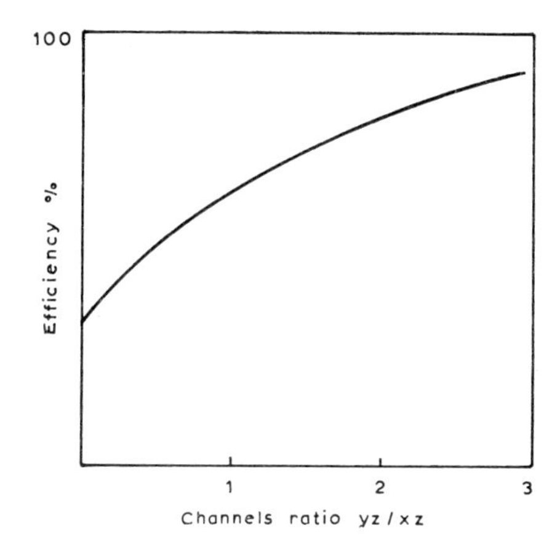

Figure 7.10. The channels ratio. The channels ratio YZ/XZ (Figure 7.8) is determined with a known amount (pCi) of the isotope in use added in a form which does not cause quenching. The efficiency is determined from the amount added and the count XZ. Successive quantities of a quenching agent such as CCl_4 are then added, the channels ratio and efficiency determined each time, and a graph plotted as in the Figure. From this curve, and a measurement of the channels ratio, the degree of quenching, and hence the efficiency, may be determined for any quantity of any preparation of the isotope in use (see Experiment 12).

with an efficiency approaching 100% in favourable circumstances. This is nearly one hundred times the efficiency of a liquid counting G-M tube. It is also possible to count neutrons. Fast neutrons can be counted using a solvent such as heptane with a high hydrogen content, and slow neutrons by loading the solvent with boron. It should thus be clear that liquid scintillation counters are versatile instruments with wide application.

7.5. *Track chambers*

These are instruments in which a charged particle produces a track which is visible and can be photographed. They are much used in nuclear physics for interpreting the results of collisions between atomic nuclei and charged particles, but they are not of much value to the radiochemist. Because of their general interest they will be briefly mentioned here.

The first instrument of this type was the Wilson cloud expansion chamber. An improved version was invented by Blackett. A very simple cloud chamber is the diffusion chamber which is continuously sensitive. Details of the construction of Blackett and diffusion chambers are given in Appendix 6. Modern applications are the spark chamber, the bubble chamber and the photographic emulsion.

7.5.1. *The spark chamber.* This consists in principle of a stack of parallel insulated plates arranged about 5 mm apart. A gas, usually neon, circulates between the plates, and the whole is surrounded with a gas-tight box. When a high energy charged particle passes through the stack, the gas in the path of the particle becomes ionized. If a high voltage is suddenly applied to alternate plates, the resultant electric field produces many more ions, and the track is revealed, in somewhat dramatic fashion, by the passage of a spark. The spark can be photographed in two mutually perpendicular directions to locate the track in three dimensions. Spark chambers are relatively simple and cheap to construct, so they can be designed for a particular experiment.

7.5.2. *The bubble chamber.* This, on the other hand, is difficult to construct and extremely costly. In principle it consists of a liquid maintained just below the boiling point, under pressure. When the pressure is slightly reduced the liquid becomes superheated at the lower pressure, and bubbles will form on the ions produced in the track of a charged particle. The advantages of the bubble chamber are that the tracks are very fine and the chamber can be filled with a pure element such as liquid hydrogen. Any events which occur can only be caused by collisions of the entering particles with a proton. Liquid helium chambers have now been constructed. The engineering and refrigeration problems of working at the very low temperature required are considerable.

Expansion cloud chambers have to be triggered before they become sensitive and then remain sensitive for a short period of time, after which they need a much longer recovery time before they can be triggered again. Spark chambers can be successfully triggered after the passage of the particle because the ions persist long enough for the triggering mechanism to operate. Bubble chambers, on the other hand, must be triggered before the passage of the particle, as the bubble forming centres caused by the particle disperse before the chamber can be expanded. This is a disadvantage. With spark chambers two detectors can be arranged on either side of the chamber in a coincidence circuit. The chamber is not triggered unless the particle passes through both detectors.

7.5.3. *The photographic emulsion.* This is really another type of track chamber which, like the diffusion chamber, is continuously sensitive. It is the simplest and cheapest of all, but has given most valuable results. Its use depends on the fact that an ionizing particle passing through the emulsion forms a latent image. By normal development the track can be seen under a microscope as a row of silver grains. Charged particles such as α-particles, protons, mesons, etc., give clear tracks, but electrons give only faint wispy lines (Experiment 34).

The emulsion used is many times thicker than a normal photographic emulsion, and thicknesses of 100 or 200 μm are common, but even so, high energy particles rarely come to rest in the emulsion and generally pass straight through. To overcome this difficulty the technique is to strip the emulsions from a number of plates and put them together like a sandwich in a light-tight package. In cosmic ray research this can be sent up in a balloon or satellite. When the package is recovered the emulsions are put back on their plates, developed, and scanned under a high power microscope. Such a technique requires a large team of workers, for each plate will require about 8 h to scan, and when an interesting event has been discovered, the various tracks in the different emulsions have to be measured, and the particles identified. This is often difficult unless the sandwich can be placed in a powerful magnetic field so that the curvature of the tracks can be used to determine the nature of the particles. No single person can undertake work of this complexity, but it is possible to obtain quite interesting results by carrying plates with nuclear emulsions up mountains. Stars produced by the disintegration of silver or bromine nuclei will be obtained above 3000 m in altitude (Figure E34 (ii)). (Details of the technique of using nuclear emulsions will be found in Experiment 34.)

8. Errors and their correction

8.1. *Introduction*

The accuracy of any quantitative experimental work is determined by the experimental error. Certain particular experimental errors occur in radioactive measurements. A clear understanding of the nature of these errors is essential to obtain an assessment of the accuracy of the result. By choosing the proper working conditions it is often possible to reduce the error of measurement until it is negligible compared to other experimental errors.

The principal sources of error are summarized as follows:

Statistical.

Background.

Resolution by the counting system.

Efficiency of the counter.

Backscatter.

Absorption,

- (*a*) by the specimen (self-absorption),
- (*b*) by the wall or window of the counter,
- (*c*) by the air.

Geometry.

Contamination of the counter.

8.2. *Statistical error*

The random nature of radioactive disintegration can be very simply demonstrated by placing a weak source of radioactivity, such as an old luminous watch, in front of a Geiger counter and counting over fixed periods of time. The precise moment at which any given atom will decay cannot be predicted, but if a large number of similar radioactive atoms is considered the laws of probability can be applied. There is no such thing as a fixed disintegration rate. There is a mean rate which theoretically can only be determined from an infinite number of observations. The probability that any given measurement of the counting rate will be within given limits of the mean rate will depend on the number of counts recorded. This probability is known as the statistical error.

The statistical error can be determined from the standard deviation (σ). This can be taken as approximately equal to the square root of the number of counts, provided this exceeds 30. It can be shown, by application of statistical theory, that there is a 68% chance that any given observation will differ from the mean count by an amount less than the standard deviation. Similarly, there is a 95% chance that the error will not be greater than twice the standard deviation. Thus with 100 counts the standard deviation is 10, but with 10 000 counts the standard deviation is 100. In the first case there is a 95% chance that the error will not be greater than 20%, but in the second case there is a similar chance that the error will not be greater than 2%. It should be clear that the largest possible number of counts should always be taken (Experiment 1).

TABLE 8.1. *Standard deviations of arithmetically combined quantities.*

Quantity	Standard deviation
A	σ_A
B	σ_B
$1/A$	σ_A/A^2
$A \pm B$	$(\sigma_A{}^2 + \sigma_B{}^2)^{1/2}$
A^2	$2\sigma_A A$
$A^{1/2}$	$\sigma_A/2A^{1/2}$
AB	$(\sigma_A{}^2 B^2 + \sigma_B{}^2 A^2)^{1/2}$
A/B	$\frac{A}{B}\left(\frac{\sigma_A{}^2}{A^2} + \frac{\sigma_B{}^2}{B^2}\right)^{1/2}$
$\ln A$	σ_A/A
$\ln(A+B)$	$\frac{(\sigma_A{}^2 + \sigma_B{}^2)^{1/2}}{A+B}$
$\ln(A/B)$	$\left(\frac{\sigma_A{}^2}{A^2} + \frac{\sigma_B{}^2}{B^2}\right)^{1/2}$
e^A	$\sigma_A e^A$

(Reproduced by permission from *Modern Radiochemical Practice* by Cook and Duncan, Oxford University Press.)

8.3. *Background*

As mentioned in Chapter 7, an end window G-M tube screened with 2·5 cm of lead will have a background of about 12 counts min^{-1}. A scintillation counter similarly screened might give one hundred or more counts min^{-1}. The background count must be determined preferably over a fairly long time and subtracted from the observed count to give the count due to the specimen.

The error due to the background depends on the difference between the background count and that of the specimen. The standard deviation on the total count of the specimen is equal to $(\sigma_b^2+\sigma_s^2)^{1/2}$ where σ_b is the standard deviation due to the background and σ_s that due to the specimen and background together. Table 8.2 shows the method of computing the error.

TABLE 8.2.

	Counts min^{-1}	Total counts	Standard deviation (σ)	σ(%)
Specimen + background	180	900	30	3·3
Background alone	20	100	10	10
Specimen alone	160	800	$(30^2+10^2)^{1/2}\simeq 32$	4

The error in this case is only slightly increased by the background. When dealing with very weak sources, however, the error due to the background may be considerable. For instance see Table 8.3.

TABLE 8.3.

	Counts min^{-1}	Total counts	Standard deviation (σ)	σ(%)
Specimen + background	40	200	14	7
Background alone	20	100	10	10
Specimen alone	20	100	$(14^2+10^2)^{1/2}\simeq 17$	17

In these cases the statistical error may be reduced by counting the background over a prolonged period. This assumes a steady background which is behaving statistically.

8.4. *Resolution by the counting system*

If the disintegration rate of the specimen is too fast for the counting system the observed count will be too low. Lost counts can be caused by failure of the G-M tube (p. 102) or scintillation counter to separate two consecutive pulses. The resolving time of most G-M tubes is about 200 μs but scintillation counters have a much shorter resolving time, about 1 or 2 μs. The counting rate may be too fast for the mechanical register, if one is fitted. Some registers will not count faster than two or three counts s^{-1}. On a simple scaler with two dekatrons before the register this gives a maximum counting rate of 18 000 min^{-1}. Losses can also occur due to failure of the electronic system to separate two pulses.

The most satisfactory way of accounting for resolution of the G-M tube or electronic system is to impose a fixed dead time with a suitable electronic circuit so that the counter is rendered inoperative for this time after each pulse. Most instruments have an adjustable dead time. 300 or 500 μs is suitable for a G-M tube. The resolving time of the electronic system is much shorter than this. The correction for lost counts is made by multiplying the dead time by the number of counts and subtracting the result from the total counting time. For example, suppose the observed counting rate were 10 000 counts min^{-1} and the dead time was fixed at 300 μs. Then the corrected time of counting would be $60\,s - (10^4 \times 3 \times 10^{-4})s = 57\,s$, and the corrected rate would be $\frac{10^4 \times 60}{57} = 10\,530$ counts min^{-1}. In this way the error due to lost counts may be eliminated.

8.5. *Efficiency of the counter*

As previously explained (p. 75), it is unnecessary in most radiochemical work to determine the absolute counting efficiency, but it is essential that the efficiency should remain constant throughout the experiment. With a G-M tube this can be achieved by maintaining the HV across the tube reasonably constant and about 50 V above the start of the plateau. With scintillation counting it is necessary to maintain more accurate control of the HV, since the plateau generally slopes much more steeply than with the G-M tube (Figure 7.6). Both G-M tubes and scintillation counters should be checked periodically with a standard source to see that their counting rate remains constant.

The most satisfactory method for determining absolute disintegration rates is to calibrate the counter with an absolute standard, which may be obtained from Amersham (Experiment 6). This standard should, if possible, be composed of the same nuclide as that used in the experiment. This will be impossible in the case of short-lived nuclides. In such cases a standard must be chosen with a similar β- or γ-spectrum. End window G-M tubes are difficult to calibrate owing to such factors as self-absorption and geometry (see following sections). Liquid counting G-M tubes of fixed capacity, and scintillation counters, can be easily calibrated using a solution of known activity. For approximate work a convenient standard for a liquid counting G-M tube is a solution of a uranium salt of known concentration. From the half-life of uranium the number of disintegrations per second of UX_2(^{234}Pa) may be determined for a known mass of the salt. (1 g of uranium in equilibrium with ^{234}Pa will give $1{\cdot}23 \times 10^4$ dis s^{-1} from the ^{234}Pa. 95% of these result in the emission of a β-particle of 2·32 MeV.) A solution of a uranium salt should be used for periodical checks on the performance of the counter, but it is only suitable as an absolute standard for use with β-emitters of nearly the same energy as ^{234}Pa.

8.6. *Backscatter*

When using end window counters a proportion of the β-particles emitted by the source will be reflected into the counter by the walls of the source holder or castle, and by the support. This reflection is called 'backscatter'. By far the greatest amount of backscatter is caused by the support. The three factors which determine the degree of backscatter are:

(1) the thickness of the support,
(2) the atomic number of the support,
(3) the energy of the radiation.

β-radiation of a given energy has a maximum depth of penetration into any given material. This depth of penetration or range can be expressed in mg cm^{-2} and is related to the energy of the radiation (Figure 9.2). The degree of backscatter increases with increasing thickness of the support and reaches a maximum at about one third of the maximum range of the β-particles in the support. This maximum thickness, expressed in the above-mentioned units, is approximately independent of atomic number, i.e. of the nature of the support, but the degree of backscatter produced increases considerably with increase in atomic number. The maximum back scattering factor varies from about 1·3 for light materials such as aluminium to 1·8 for dense materials of high atomic number such as lead or platinum. This factor is nearly independent of the energy of the β-particles above about 0·6 MeV.

Thus the efficiency of an end window G-M counter will vary considerably with the nature and thickness of the support for the specimen, and it will also be affected by the type of material used for the housing. To obtain reproducible results when using an end window counter it is essential to use the same thickness and material for the support throughout the experiment.

8.7. *Absorption*

α- and β-particles are absorbed within the specimen. Different thicknesses of the same material will therefore give different counting rates. When preparing solid specimens for counting it is essential to use the same mass of material, or a specimen of 'infinite thickness' (see below), spread over the same area in order to obtain reproducible results. The degree of absorption depends on the energy of the α- or β-particles. With α-particles and low energy β-particles a source must be extremely thin to avoid errors due to self absorption. The error can be avoided by using a specimen of infinite thickness, i.e. of such a thickness that none of the particles from the lowest layer enter the counter. This technique is particularly useful when counting solid specimens containing ^{14}C or ^{35}S. The count rate obtained will then depend only on the area and specific activity of the material.

With liquid counting G-M tubes which have a constant wall thickness, errors can be introduced by using liquids of different density since the self absorption increases with increase in density. For example, suppose a specimen of iodoethane contained both free and combined radioactive iodine. The proportion of free to combined radioactive iodine could be determined by counting an aliquot of the iodoethane before and after extraction with an aqueous solution of sodium sulphite. If the aqueous layer was counted instead of the organic layer, too high a result would be obtained as the density of the organic layer is nearly twice the density of the aqueous layer.

Absorption losses also occur in the wall of a liquid counter or the window of an end-window counter, but such losses do not affect comparative counting, though they do affect counter efficiency. For counting solids which are α-emitters or soft β-emitters such as ^{14}C and ^{35}S it is essential to use an end window counter with a window thickness of less than 2 mg cm^{-2}. Even in this case losses of up to 90% will result with ^{14}C and ^{35}S. Considerable losses will also result from absorption by the air and the specimens must be placed as close to the counter window as practicable. Even so the overall efficiency is unlikely to exceed 1%.

8.8. *Geometry*

To obtain reproducible results it is essential to place the source to be counted in exactly the same position with respect to the counter. This is not easy when using an end window counter. Because of this the liquid counting G-M tube, the well type or annular type scintillation counter, or the liquid scintillation counter are much to be preferred for radiochemical work.

8.9. *Contamination of the counter*

Liquid counting G-M tubes, the plastic cups used for annular scintillation counters, the specimen tubes for well-type scintillation counters, and the silica vials of liquid scintillation counters must be decontaminated between each count. The plastic cups and specimen tubes present few problems. They cost a few pence and if seriously contaminated with long lived radioactive isotopes they can be discarded. A number can be used during a particular experiment and at the end they can be washed with carrier and left to soak, if necessary, in a beaker of detergent such as Decon. Specimen tubes should *always* be checked for cracks before use.

Liquid counting G-M tubes, on the other hand, must be decontaminated during the course of the experiment. Washing with carrier will often be sufficient though it may be necessary to use a detergent. Liquid counting tubes should not be left filled with an alkali detergent for any length of time owing to attack on the thin glass envelope by the alkali.

The silica vials of liquid scintillation counters must not be decontaminated with an alkali detergent as this will creep over the surface and attack the aluminium reflector. They should be left overnight filled with a non-alkaline detergent after thorough rinsing with carrier.

All equipment used for counting liquids must be checked for background after decontamination.

9. Energy determination

9.1. *Introduction*

The energy of the radiation emitted by a radioactive nuclide is a characteristic of that nuclide. Measurement of the energy is often a valuable means of identification. It is also necessary to know the energy in order to determine the amount of screening required for safe working. The methods used in practice depend on the type of radiation.

For α- and β-particles the simplest method is to determine the thickness of an absorber which is required to stop the particles completely. This thickness is known as the range. To a first approximation the stopping power of different materials is nearly independent of atomic number and is a function of the mass interposed in the path of the radiation. It is conveniently measured in mg cm^{-2}.

9.2. *Alpha particles*

The most satisfactory medium is air. The source is placed in front of the detector and the beam of α-particles suitably collimated. The source is slowly withdrawn until the counting rate falls to the background. The range in mg cm^{-2} is determined by measuring the distance from the source to the counter and taking the density of air as 1·29 kg m^{-3}. A correction must be added for the thickness of the counter window. The most suitable detector is a solid state detector (p. 104) since, by use of a discriminator, it can be adjusted to count α-particles only, in the presence of both β-particles and γ-rays. An improved method is to mount the source and the detector at the opposite ends of a glass tube which is connected to a vacuum pump and manometer, and to reduce the pressure until the detector begins to operate. The density of the air in the tube can be determined from the pressure and the temperature. Thus the range of the α-particles in air at STP may be calculated. This method is similar to that used by Rutherford (p. 10) except that a solid state detector is used instead of a fluorescent screen. If it is necessary to determine the energy of the α-particles this can be found from the relationship $R = 0{\cdot}306\ E^{3/2}$ where R is the range in cm of air at STP and E is the energy in MeV. The range of α-particles of different energies may be compared by measuring the length of the tracks produced in a photographic emulsion (Experiment 34 (*a*)).

9.3. *β-particles*

To make an absolute measurement of β-ray energy it is necessary to use a β-ray spectrometer, but approximate values of maximum β-ray energy may be obtained by plotting absorption curves. The absorber normally used is aluminium. The source is arranged at a fixed distance from the window of a G-M tube and sheets of aluminium of known thickness, expressed in mg cm^{-2}, are interposed between the source and the tube, as close to the tube as possible. The activity, expressed in counts min^{-1}, is measured with increasing thicknesses of aluminium and the log of the counting rate is plotted

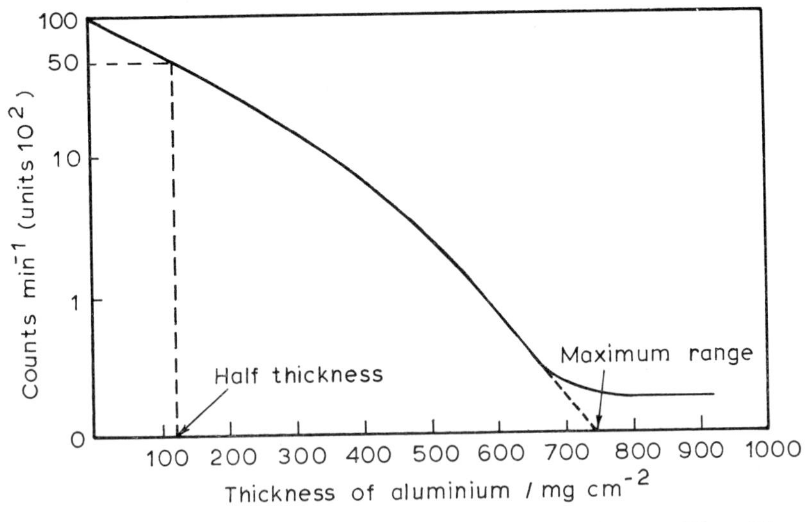

Figure 9.1. To determine the maximum β-energy of a nuclide. The maximum energy is determined from the maximum range, or the half-thickness, and the use of Figures 9.2 or 9.3. The exponential nature of the first part of the graph is a fortuitous result of the continuous β-spectrum and the effect of backscatter. The graph eventually becomes horizontal, somewhat above the background, due to Bremsstrahlung: X-rays produced by the slowing down of the β-particles in matter. Bremsstrahlung intensity is proportional to Z^2, hence the use of aluminium rather than lead to measure maximum energy (see Experiment 10).

against the thickness. The resulting graph is nearly linear (Figure 9.1). The exponential absorption of β-radiation from a simple β-emitter is a fortuitous result of the continuous β-spectrum and the effects of scattering. The curve eventually flattens out to give a steady counting rate which is somewhat above the background. One reason for this is that electrons are scattered out of the aluminium, and the expulsion of these electrons produces secondary X-rays. This scattered radiation is known as 'Bremsstrahlung'. Another cause of the increased count above the background is γ-radiation emitted by the source, since the attenuation of γ-radiation by aluminium foil is slight; pure β-emitters give the most satisfactory absorption curves. The range of the β-particles is determined from the graph, and the maximum β-energy by comparison with a range energy curve (Figure 9.2). The main disadvantage of this method is that the statistical error becomes high as the curve flattens and this causes inaccuracy in determining the range.†

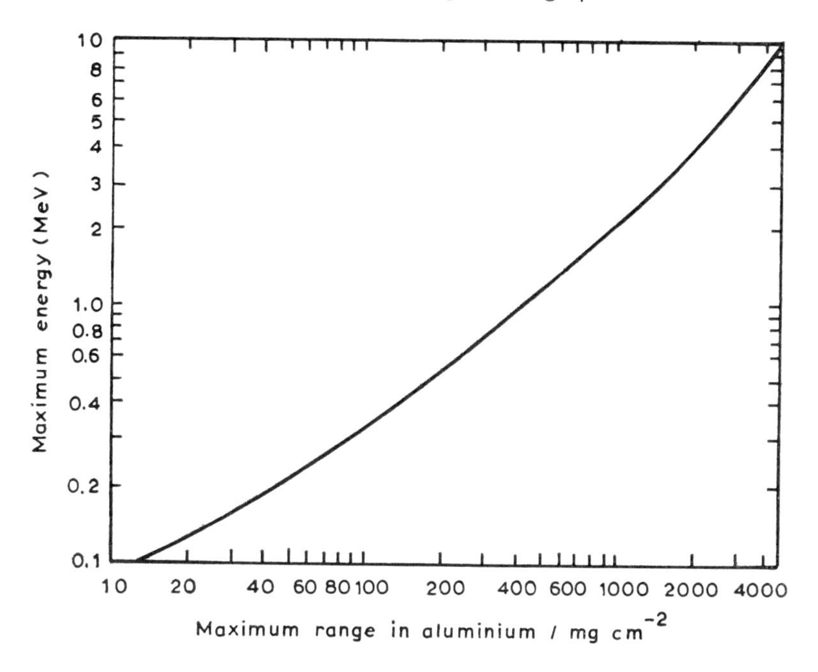

Figure 9.2. Maximum energy of β-particles against range in aluminium. The graph is used to determine the maximum energy of β-particles emitted by a nuclide following measurements as shown in Figure 9.1. The determination of maximum energy is a useful means of identification. (From Katz and Penfold, *Rev. Mod. Phys.*, **24,** 28, 1952.)

† Allowance should be made for the thickness of the counter window.

A more convenient method of determining the maximum energy of β-particles is to measure the 'half thickness', which is the thickness of aluminium required to reduce the counting rate to half its initial value. The energy is determined by comparison with a graph of half thickness against maximum β-energy (Figure 9.3) (Experiment 10).

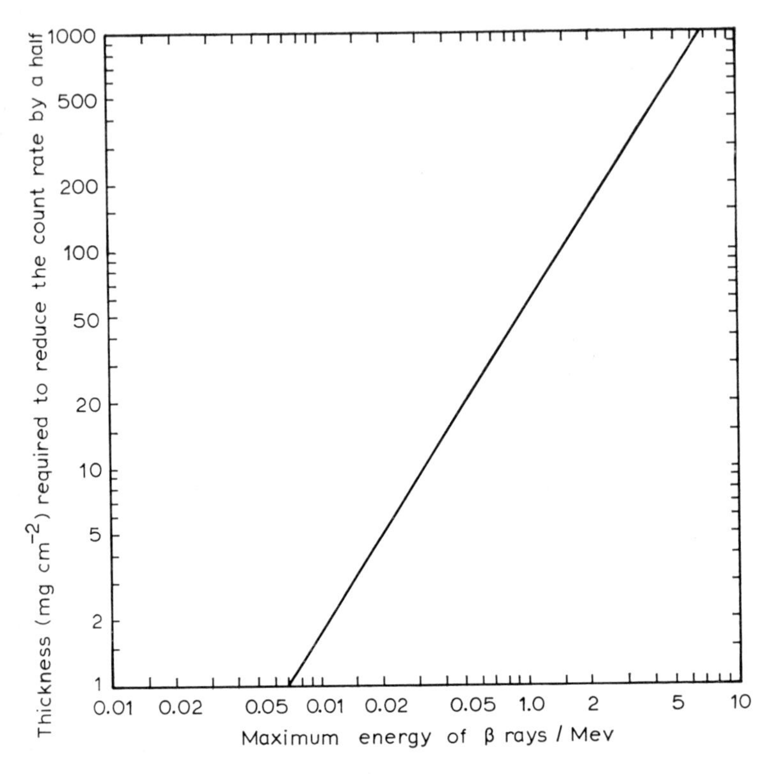

Figure 9.3. Maximum energy of β-particles against half-thickness of aluminium. This is a quicker and more convenient method of determining maximum energy, particularly for high energy β-emitters.

9.4. *Complex absorption curves*

Absorption curves frequently show points of inflection. These are caused by the presence of different components, such as a mixture of two β-emitters. By resolution of these curves it is possible to identify the two nuclides from their maximum β-energy [Figure 9.4 and Experiment 24 (*b*) (ii)]. Unless the slopes differ by a factor of at least 2 it is very difficult to resolve the components. With high energy β-particles which require a considerable thickness

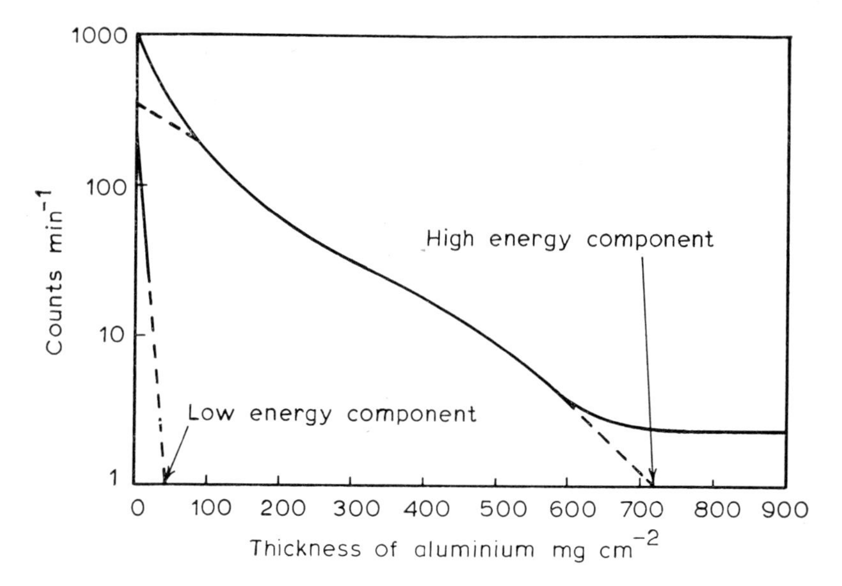

Figure 9.4. Resolution of a complex β-absorption curve. The graph was obtained in the author's laboratory from a mixture of ^{35}S and ^{32}P. The maximum energy of the low energy component is determined by subtracting the extrapolated portion of the high energy component from the mixture as given by the upper curve (see Experiment 24 (*b*)).

of aluminium to stop them completely, a point of inflection may also be caused by the presence of a low energy gamma component. If the activity due to the γ-rays is extrapolated back to zero absorption, it can be subtracted from the main curve to obtain the contribution of the γ-rays at any given thickness.

9.5. *Semiconductors*

As indicated on p. 106 a semiconductor with a suitable amplifier and pulse height analyser can be used to measure the energy of both α- and β-particles (p. 107).

9.6. *Gamma rays*

In principle it is possible to determine γ-ray energy by absorption measurements similar to those used for β-particle energy, but lead is normally used in preference to aluminium owing to the greater density. The half-thickness method is employed, and the γ-ray energy determined from a graph. The method is much less accurate for γ-rays than for beta particles, owing to scattering effects produced by interaction of the γ-rays with materials surrounding the counter. In addition, high energy electrons are scattered out of the lead. The best results are obtained by placing the source at a considerable distance from the counter, collimating the γ-rays, and inserting the lead absorbers mid-way between the source and the counter. In addition a thick sheet of aluminium should be placed in front of the counter window to absorb high energy electrons scattered out of the lead.

9.7. *The γ-ray spectrometer*

Though absorption measurements are useful to determine the thickness required to reduce the γ-radiation from a particular nuclide to safe proportions, it is necessary to use a γ-ray spectrometer to determine the energy precisely. For this purpose a sodium iodide crystal is required in conjunction with a photomultiplier tube (p. 108). The electronic equipment needed consists of a linear amplifier, a pulse height analyser and a ratemeter or scaler. Some form of recorder is a great advantage.

γ-rays or photons are of the same nature as light photons, but of higher energy. Light photons emitted in the middle of the visible spectrum have an energy of about 2 eV, whereas a typical γ-photon emitted in a radioactive disintegration may have an energy of 1 MeV, equivalent to the energy of 500 000 visible light photons. The function of the scintillator is to convert as much of the energy of the γ-photons as possible into light photons. Unfortunately, this process is very inefficient. The best scintillator is sodium iodide activated with a trace of thallium, but the efficiency is never greater than 10%. As explained on p. 107 the scintillations produced are very faint and must be observed with a photomultiplier tube. The action of the photomultiplier has already been explained (p. 107), but the appearance of the actual spectrum produced will now be discussed in detail.

Every γ-ray belonging to a particular energy group emitted by a given nuclide has exactly the same energy and so will have a sharp line spectrum. In practice, however, a γ-spectrometer does not produce a line spectrum but a series of peaks, the centre of the peak corresponding to the energy of the γ-ray, and the base of varying width according to the quality of the crystal and photomultiplier tube. This peak broadening is caused by two sets of factors. The first concerns the statistical nature of the energy conversion processes in the crystal and photomultiplier tube over which there is no control. The second is due to variations in the light transmission in different parts of the crystal, to inefficient electron collection by the dynodes in the photomultiplier tube, and to irregularities in the sensitivity of the photo-cathodes. Good manufacturing techniques, careful selection of crystals, and the surrounding of the crystal with a good reflector will reduce the second set of factors to a minimum. It is necessary to ensure good optical coupling of the crystal to the photomultiplier tube by using a silicone oil. (p. 108).

In addition to the peak which corresponds to the total energy of the γ-ray, lower energy components are also found (Figure 9.5). These are caused by the various interactions occurring in the crystal. A γ-ray entering the crystal may be photoelectrically absorbed (p. 109). The high energy electron released interacts strongly with the crystal lattice. A fraction of the energy thus released (about 5–10%, as previously mentioned), is converted into visible light photons which produce the main or total energy peak. In addition to photoelectric absorption, Compton scattering occurs (p. 93). γ-rays will undergo deflections through varying angles, and recoil electrons of different energies are produced which give rise to the lower energy components in the spectrum. These lower energy components reach a maximum at 180° scattering, which corresponds to the Compton edge (Figure 9.5). The scattered γ-rays may pass right out of the crystal, may undergo multiple scattering, or may finally be photoelectrically absorbed. If this latter process occurs, the photomultiplier tube is unable to resolve the series of scintillations produced by multiple scattering and photoelectric absorption, since they occur so rapidly, and a single voltage pulse is recorded, with an energy equal to that produced by a single photoelectric scintillation; this pulse is added to the main photoelectric peak. The bigger the crystal the better the chance that photoelectric absorption will occur, and the larger will be the photoelectric peak in comparison to the lower energy components. Another peak is generally found in the low energy region of the spectrum. This is caused by photoelectric absorption of degraded γ-rays scattered from the sides of the crystal or reflected off the front face (Figure 9.5).

Frequently a nuclide gives rise to more than one γ-ray following disintegration. ^{60}Co is a good example. This emits two γ-rays of energy 1·17 and 1·33 MeV respectively. Usually only one of these is captured at a time, the other missing the crystal or passing through it. The resultant spectrum contains two peaks corresponding to the energies of the two γ-rays. The first peak will be higher than the second due to the Compton edge of the second peak being superimposed on the first peak (Figure 9.6 and Experiment 9). Occasionally both γ-rays will be captured simultaneously and this will produce a small peak corresponding to an energy of 2·5 MeV.

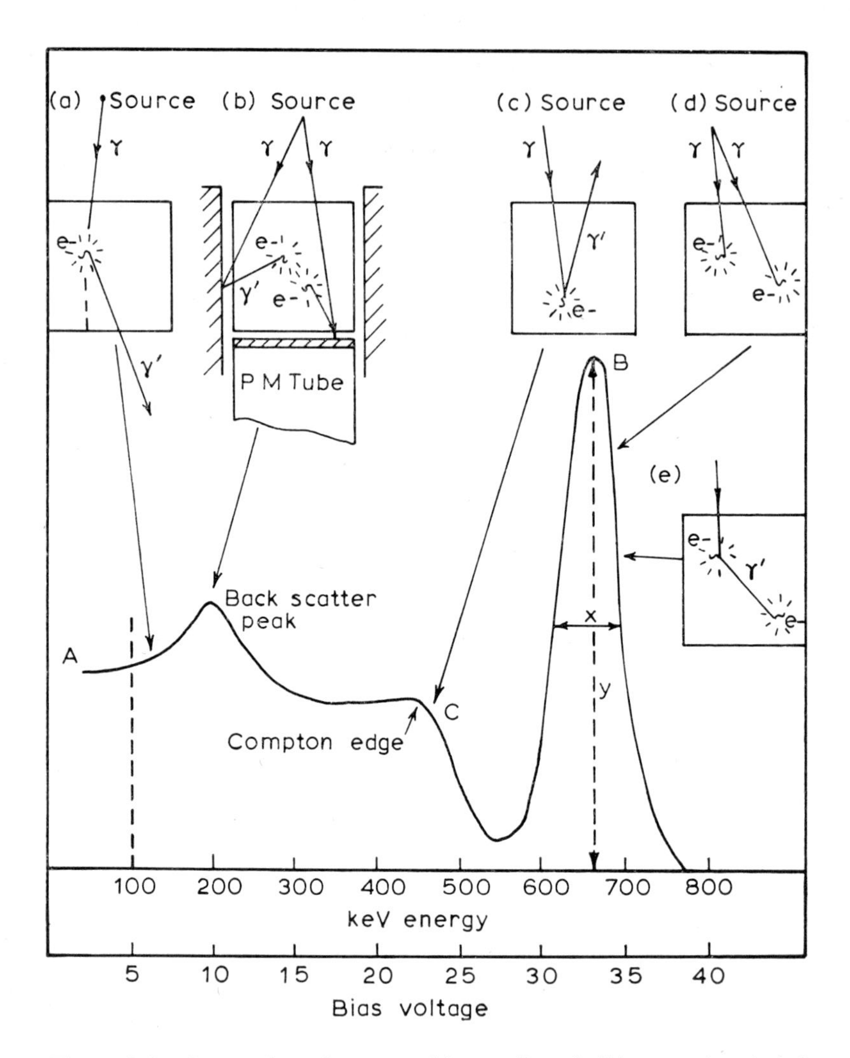

Figure 9.5. Interaction of γ-rays with a sodium iodide crystal and their contribution to the γ-spectrum of ^{137}Cs. e^- shows an electron ejected by a γ-ray. The electron interacts with the phosphor to produce a light photon, the energy of the photon being proportional to the energy given to the electron by the γ-ray. (*a*), (*b*) and (*c*) show various types of scatter which all contribute to the spectrum from A to C. (*d*) and (*e*) show photoelectric absorption events, (*e*) showing scatter followed by photoelectric absorption. Both such events contribute to the photopeak B. γ' indicates a scattered γ-ray. (See Experiment 9). The resolution of the spectrometer may be expressed as $(x/y) \times 100$. (P. 135).

With γ-rays of high energy, pair production may result (p. 93) and this will produce a peak at 1·02 MeV. Any γ-ray with an energy greater than this value may create an electron–positron pair, but the number of these events is small except with very high energy γ-rays.

9.8. *Measurement of resolution*

The resolution of a γ-ray spectrometer is a measure of the separation of two γ-rays which are close together in energy. It depends on the peak width already discussed. There are two convenient methods. One uses a nuclide such as ^{137}Cs, which has a single line in the γ-ray spectrum at 0·662 MeV. The resolution is expressed as a percentage of the peak width at half height divided by the height of the peak above the background. A value below 8 % for a sodium iodide crystal is considered to be very good (Figure 9·5).

Another method is to use the two peaks of ^{60}Co, and to divide the height of the higher energy peak by the height of the dip between the peaks. Any ratio above 3 : 1 is good (Figure 9.6).

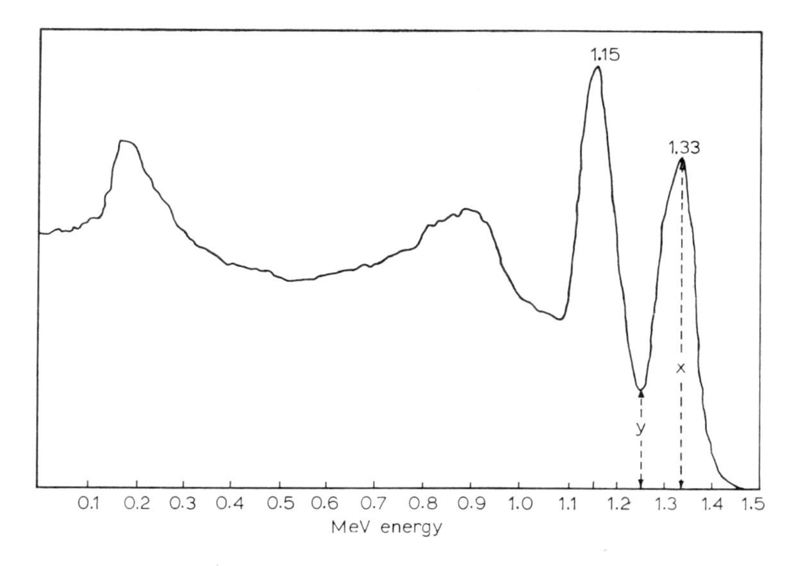

Figure 9.6. The γ-spectrum of ^{60}Co. Two γ-rays are emitted of energy 1·15 and 1·33 MeV respectively. The height of the first peak is greater than that of the second as the Compton scatter associated with the second peak is superimposed on the first. The resolution of the spectrometer may also be measured from this graph and expressed as $x/y = 10{\cdot}5/3{\cdot}2 = 3{\cdot}3$, which is good for a sodium iodide crystal. (From a recording made in the author's laboratory, see Experiment 9).

9.9. *Calibration*

A γ-ray spectrometer may be calibrated by using a number of nuclides whose γ-energies are accurately known, such as ^{137}Cs (0·662 MeV), ^{131}I (0·364 MeV), ^{198}Au (0·400 MeV), ^{144}Pr (0·133 MeV), ^{60}Co (1·17 and 1·33 MeV respectively) and plotting the energy against the meter reading on the pulse height dial. The plot obtained should be linear. It is important to keep the HV constant throughout and to set the gate width in the pulse height analyser to a fixed value. The gain of the amplifier must also be fixed. The position of the peaks is very sensitive to slight changes in HV. Increase in HV moves the peaks higher on the pulse height dial. Doubling the amplifier gain should double the meter reading fro a given photon peak. By maintaining the HV fixed and increasing the amplifier gain, it is possible to obtain more detail in the different parts of the spectrum and also to detect low energy peaks which would be below the threshold of the gating circuit.

The gate width setting will depend on the intensity of the available sources. Reducing the gate width improves the resolution, provided the source is of sufficient intensity. Some γ-ray spectrometers have a percentage scale on the selector switch. This causes the gate width to vary automatically as a percentage of the threshold level, according to the pulse height, reaching full width at the end of the scale. Such a device is very useful when plotting a spectrum consisting of both low and high energy γ-rays, since the low energy γ-rays are more efficiently absorbed and so produce much larger peaks than the high energy γ-rays.

γ-ray spectrometers are very useful in analysis, as it is possible to use them to determine several different γ-emitting isotopes in the same material without doing a chemical separation. The intensity of the respective photon peaks is measured, and a correction made for the Compton scattering associated with the high energy photon peaks which is superimposed on the low energy photon peaks. Certain alloys may be analysed quantitatively by neutron activation in a reactor followed by counting with a γ-ray spectrometer. It is, of course, necessary to know what metals are present, the capture cross-sections of their nuclei, and the half-lives of the isotopes produced (Experiment 14).

The most recent γ-ray spectrometers use a large solid-state detector, maintained permanently at the temperature of liquid nitrogen, in conjunction with a multichannel analyser. The solid-state detector gives much better resolution of γ-ray energies than a sodium iodide crystal, the peaks hardly showing any broadening, but the sensitivity is low. The equipment is extremely costly (see p. 107).

10. Some uses of tracers in chemistry

10.1. *Introduction*

Radioactive isotopes have wide applications both in research and teaching. In some cases they make experiments possible which could not be carried out in any other way. In others they offer a more convenient and a simpler procedure than the conventional method.

Before carrying out an experiment using a radioactive tracer it is essential to be thoroughly familiar with the experimental technique. If there is any doubt a ' cold ' experiment should first be conducted, i.e. without the use of a tracer.

The basic assumption in all tracer work is that the radioactive isotope behaves chemically in an identical manner to a non-radioactive isotope. This assumption is valid in the great majority of cases, though hydrogen and some of the lighter elements show some departures from this rule, owing to large differences in relative atomic mass (atomic weight). The average kinetic energies of ^{1}H and ^{3}H will be identical at the same temperature, but the velocities will be in the ratio of 1·73 : 1. This would make a considerable difference in kinetic studies involving ions, but little difference if the ^{3}H were combined in a compound.

10.2. *Factors to be considered when the use of a tracer is contemplated*

(1) The availability of a suitable tracer.

(2) The detection equipment obtainable.

(3) The hazard.

(4) The presence or absence of daughter products.

(5) The chemical form in which the tracer is required.

10.2.1. *Availability.* Radioisotopes of convenient half-life are now available for nearly all the known elements. Notable exceptions are lithium, beryllium, boron, nitrogen and oxygen, either because the radio-isotopes have too short half-lives, or because the capture cross-section is too low and the half-life too long for convenient production, as in the case of ^{10}Be. Radioisotopes which can be prepared by (n, γ) reaction, or as separated fission products, may be purchased comparatively cheaply. Those which have to be made by charged particle bombardment are much more expensive. A few useful radioisotopes may be prepared by irradiation of a suitable compound with a laboratory neutron source and the isotope separated by a recoil process in high specific activity (see p. 149). Examples are ^{82}Br ($t_{1/2}$ = 36 h), ^{128}I ($t_{1/2}$ = 25 min), ^{56}Mn ($t_{1/2}$ = 2·6 h). There are two useful naturally occurring radioisotopes, ^{212}Pb ($t_{1/2}$ = 10·6 h), ^{212}Bi ($t_{1/2}$ = 1 h). Both may be easily separated from a thorium compound (Experiments 2, 17 and 18).

10.2.2. *Detection equipment.* The three types of detection equipment most commonly used in tracer work are:

(1) liquid counting G-M tubes for medium to high energy β-particles;

(2) sodium iodide scintillation counters, or solid state detectors for γ-rays;

(3) liquid scintillation counters for low energy β-particles.

The operation of these detectors has already been covered in Chapter 7. Here their main practical advantages and disadvantages will be briefly discussed.

Liquid counting G-M tubes have the advantage that they can be operated with simple and relatively cheap ancillary equipment. They have reasonably good detection coefficients (5–10% for β-particles with energies in excess of 1·5 MeV). They have a low background count. However, they must be decontaminated and checked between each observation and this is tedious and time consuming.

Sodium iodide scintillation counters using a 'well-type' crystal are much more sensitive and convenient than G-M tubes for counting radioisotopes which emit γ-rays. Test-tubes holding up to 4 cm^3 may be used and these can be decontaminated after the experiment. Their efficiency for ^{131}I or ^{110m}Ag is 30–40% which may be compared with 0·5% for a G-M tube (see p. 168 for their use in analysis).

Liquid scintillation counters are particularly useful for counting ^{14}C, ^{35}S and ^{3}H owing to their high efficiencies. The particular problems associated with their use have been fully discussed (p. 112). Their main disadvantage is the higher cost of the operating equipment.

10.2.3. *The hazard.* Considerations of radiological safety are of paramount importance. The hazard depends on three factors:

(1) the nature of the isotope
(2) the quantity required
(3) the chemical form of the isotope.

(1) Radioactive isotopes are classified into three main groups: high, medium and low toxicity. The maximum permissible yearly intake for the majority has been fixed by international agreement. Table 12.2, p. 163, gives the value for a number in common use. (For further details the reader should consult Chapter 12.)

(2) The quantity required will depend on the number of experiments to be carried out, the efficiency of the detection equipment, the statistical accuracy required, the time available for counting and the specific activity of the isotope supplied. These factors must all be considered in relation to (1) before deciding on the feasibility of the experiment. The reader is referred to Question 10.1, which deals with a problem of this type.

(3) Isotopes which are combined in the form of volatile liquids or gases, or are present as the free element, should be handled in a glove box or fume chamber, unless the quantity is well below the maximum permissible yearly intake.

10.2.4. *Daughter products.* Radioactive isotopes frequently give rise to radioactive daughter products. Typical examples are $^{137}Cs \rightarrow ^{137m}Ba$, $^{144}Ce \rightarrow ^{144}Pr$, $^{90}Sr \rightarrow ^{90}Y$, $^{132}Te \rightarrow ^{132}I$, $^{212}Pb \rightarrow ^{212}Bi$. These cases all show transient equilibrium (p. 70), since the daughter has a shorter half-life than the parent. If detection is to be carried out through the radiation from the daughter, it is essential to count samples which have reached transient equilibrium (Experiments 17 and 18).

10.2.5. *Chemical form.* Radioactive isotopes are generally supplied as ions of the free element, or in some form of chemical combination. If a solution is kept for a few days chemical changes may occur. A good example of this is ^{131}I, which is commonly supplied as I^-. Rapid oxidation of I^- occurs, due to radiolysis of the water, forming IO^- or IO_3^-:

$$H_2O + \gamma \rightarrow HO^\cdot + H^\cdot$$
$$HO^\cdot + HO^\cdot \rightarrow H_2O_2$$

Samples of ^{131}I are often stabilized by adding a trace of thiosulphate as a reducing agent.

^{14}C, when supplied as $NaH^{14}CO_3$, exchanges with CO_2 in the air. Unless the solid or solution is kept tightly corked the activity rapidly disappears.

In many tracer experiments the radioactive isotope may be required in a different chemical form to that in which it is supplied. Chemical reactions involving synthesis or exchange must then be conducted (see below).

10.3. *Applications to chemical processes*

10.3.1. *Analysis.* (*a*) Neutron activation. This method can be used both for qualitative and quantitative analysis. Provided the elements present can be activated to produce γ-emitters they may be detected by the energy peaks. By following the decay of these peaks the half-lives may be determined. These two parameters are generally sufficient to identify the isotope and no chemical separation is necessary. The technique is particularly useful for valuable archeological specimens (Experiment 14 (*a*) is an example of qualitative analysis).

For quantitative work it is necessary to irradiate a known mass of the element, or of a pure compound, in a known neutron flux for a known time, and to compare the activity with that produced by irradiation of the unknown under similar conditions. Corrections for decay and for the formation of daughter products must be made. The activity produced will depend on the mass present, the neutron flux, the capture cross-section, the time of irradiation and the half-life (unless the specimens are irradiated to saturation) (Experiment 14b). For pure β-emitters the technique is not so simple. It is necessary to use either a thin specimen of the same thickness for both known and unknown, or to use infinitely thick specimens.† If two or more activities are present it is generally necessary to carry out a chemical separation, though two different β-activities may be assayed simultaneously using liquid scintillation counting (p. 188).

A recent application of neutron activation was the analysis of lunar specimens. In a study of sample 12 070 no less than 62 of the 72 elements present were determined by this method [37].

Portable neutron sources may be used for neutron activation provided the capture cross-section of the element to be assayed is not too low. ^{241}Am–Be sources are convenient as they have a low γ-background and the neutron output is $2 \times 10^6\,n\,s^{-1}\,Ci^{-1}$. The neutrons are of high energy and the source must be kept in a suitable

† This applies to solid insoluble substances. If the specimen could be dissolved in a suitable solvent, a liquid counting G-M tube could be used.

moderator, such as a cylinder of paraffin wax large enough to reduce the high energy neutrons to thermal energies. The long half-life of ^{241}Am (433 y) is an additional advantage. ^{124}Sb–Be sources produce low energy neutrons, but the γ-background is high and considerable shielding is required. The half-life of 60 d is also inconvenient. ^{252}Cf, as mentioned on p. 54, is now available as a neutron source. This isotope decays mainly by α-emission, but 3·1% of the decay events are spontaneous fissions with an average neutron yield of about 4 per fission. The half-life for this decay is 2·65 y. The main advantage of this source is the high neutron output (4×10^8 n s^{-1} Ci^{-1}) and the cost is considerably lower than other neutron sources of comparable output. The γ-background is high and considerable shielding is required, and at least 50 cm of water, polythene or paraffin wax to moderate the 7 MeV neutrons.

Neutron sources are valuable for 'on the spot' neutron activation analysis, and it is possible to assay short lived products by the prompt γ-rays emitted during the neutron activation, thus avoiding loss of activity through decay following separation.

(*b*) Isotopic dilution. This method is particularly valuable when it is necessary to assay a compound in a mixture from which it cannot be easily separated quantitatively. A known mass (y) of the compound containing a radioactive isotope of known specific activity (S_1) is added to a known mass of the mixture, and after thorough mixing some of the compound is separated and its specific activity determined (S_2).

Let x be the mass of the compound present in the known mass of the mixture. Then:

$$(x+y)S_2 = yS_1$$

and

$$x = y(S_1 - S_2)/S_2 \tag{10.1}$$

The method may be applied in reverse by adding carrier-free tracer, separating some of the compound and measuring the specific activity. This sample is then diluted with a known mass of inactive compound and the specific activity again measured. The sensitivity of the method will depend on the activity of the tracer added.

Isotopic dilution analysis has advantages over neutron activation analysis if there is interference from γ-spectra of other isotopes produced by the neutron activation. The method is particularly useful for the determination of impurities in the atomic number range 35–70 present in fissile material, where neutron activation would confuse the issue by the production of fission products of the same elements. Isotopic dilution analysis is used in diagnostic medicine to determine total blood volume by injection of a known quantity of a labelled sodium or iodine compound into the blood stream and withdrawing a sample of blood after complete mixing has been achieved.

(*c*) Radioisotopic exchange. The method makes use of the fact that when exchange occurs between a labelled element in a compound and another inactive sample of the element in a different chemical form, the specific activity at equilibrium is the same in both samples. An example of this method is the determination of an organic iodide in a mixture by heating the mixture with labelled inorganic iodide, of known specific activity, dissolved in a suitable solvent. When equilibrium has been reached the inorganic iodide is extracted and its residual specific activity determined. This is then compared with the activity of the organic mixture.

(*d*) Radiometric titration. Titration methods can be used provided phase separation occurs, such as the formation of a precipitate (see Experiment 23); solvent extraction is an alternative. A recent application is the determination of very low concentrations of iodide in aqueous solution using Hg(II) labelled with ^{203}Hg. The precipitate of HgI_2 is extracted into methylbenzene. The end point of the titration is the presence of non-extractable activity in the aqueous phase. The limit of iodide determination is reported as 10^{-9} g [38].

10.3.2. *Exchange reactions.* This type of reaction involving the same element in different chemical states can only be studied with the aid of isotopes and radioactive isotopes are ideal for the purpose. As exchange may often be easily demonstrated some experimental procedures will be briefly described.

One of the simplest examples is the exchange between ions in a precipitate and the same ions in solution. G. Hevesy, a pioneer of radiochemistry, showed in 1920 that exchange occurred between lead ions in precipitated lead chloride and lead ions in solution [39]; he used ^{212}Pb, obtained from thorium, as a tracer. A good demonstration of this type of exchange is to use silver chloride labelled with ^{110m}Ag. The precipitate must be washed, half of it shaken with pure water, and the other half with a concentrated solution of silver nitrate. If a scintillation counter is used 0·5 μCi of ^{110m}Ag is quite sufficient for the experiment, but about 5 μCi should be used with a G-M tube. No effect will be observed in the water unless the specific activity of the AgCl is very high, but a considerable count will be observed in the silver nitrate. It is important to use a freshly prepared precipitate and to protect it from light. AgCl which has darkened on exposure to light exchanges very slowly. Ten minutes shaking or stirring is sufficient.

Another exchange reaction of considerable interest is that between a metal and ions in solution. Zinc is a suitable metal. ^{65}Zn ($t_{1/2}=245$ d) may be produced with a specific activity of 1 mCi g^{-1} by irradiation in a flux of 10^{12} n cm^{-2} s^{-1} for 7 days. 100 μCi will be sufficient for repeated use over several years. Care must be taken to adjust the pH of the zinc salt solution to 7, to prevent dissolution of the zinc. Another suitable metal is silver. A clean well polished piece of silver foil should be immersed in silver nitrate solution labelled with ^{110m}Ag. At room temperature appreciable exchange will take place in 15 min. In both of these cases it is instructive to plot the activity produced against log t, since the rate of diffusion into the solid is an exponential function of the thickness.

Exchange of halogen atoms with halide ions is extremely rapid and may be demonstrated by shaking a solution of iodine or bromine in an organic solvent with an aqueous solution of labelled iodide or bromide. If a neutron source is available ^{128}I or ^{82}Br may be prepared by Szilard–Chalmers reaction (p. 148), otherwise ^{131}I or ^{82}Br must be purchased. The aqueous layer will become coloured and the organic layer will become active. The equilibrium is dynamic. Halogen molecules are passing both ways across the phase boundary. In the aqueous layer they react with halogen ions to form I_3^- and Br_3^- respectively, and some of these complexes are labelled. On dissociation either of the bonds in the linear complex may break so that some halogen molecules become labelled and pass into the organic layer. For success with this experiment it is important that the specific activity of the aqueous layer should not be too low. 0·1 μCi of iodide or bromide in 25–50 cm^3 of 0·1 M potassium iodide or bromide will give good results.

A suitable exchange reaction for quantitative study is that between halide ions and an organic halide in alcoholic solution (see Experiments 25 and 26).

Electron exchange or redox reactions may be followed using a radioactive tracer. One such reaction for which a tracer is essential is the electron exchange between two different oxidation states of the same element, such as the equilibrium reaction between MnO_4^- and MnO_4^{2-} or Fe^{3+} and Fe^{2+}. Both of these exchange reactions are extremely rapid, but the oxidation of Fe^{2+} by Fe^{3+} may be shown by using labelled Fe^{2+}, carrying out the reaction in 5 or 6 molar hydrochloric acid and extracting the Fe^{3+} as iron(III) chloride, with ether.

The non-equivalence of the two sulphur atoms in $S_2O_3^{2-}$ may be demonstrated by boiling sodium sulphite solution with labelled sulphur to produce thiosulphate, and then decomposing the thiosulphate with acid. The whole of the activity may be recovered in the precipitated sulphur. Labelled sulphur may be prepared by reducing labelled sodium sulphate with heated charcoal.

10.3.3. *Diffusion.* This process may be simply demonstrated and some practical illustrations will be briefly described. One such process is the diffusion of metals into metals. An interesting example is lead into lead. A piece of clean lead sheet should be placed in a concentrated solution of a thorium salt. The ^{212}Pb present as ions in the solution will exchange with inactive lead, and the lead sheet will become active. After thorough washing and drying the lead sheet should be pressed into contact with a thin lead foil of sufficient thickness to absorb all the α-particles given off by the daughter products of ^{212}Pb (^{212}Bi and ^{212}Po). After several hours an α-activity will be detected on the side of the foil opposite to the lead sheet.

Another interesting case of diffusion for which an isotope is essential is self-diffusion in a salt solution. Two identical solutions, one containing a tracer, should be separated with a sintered glass partition. The rate of diffusion of the two ions of a salt solution may be determined independently, e.g. sodium chloride labelled with ^{24}Na and ^{36}Cl or ^{38}Cl.

Biological studies such as the rate of circulation of sodium ions *in vivo* may be followed by injecting radioactive saline. The rate of uptake of phosphate or sulphate ions in plants is easily determined and the distribution of the phosphate or sulphate in the plant may be detected by autoradiography (Experiment 33).

10.3.4. *Complex ions.* The nature of the charge carried by complex ions may be investigated by electromigration experiments using sintered glass partitions. An example is the change in the sign of the charge carried by complex nitrates of the lanthanide ions in varying concentrations of nitric acid. Labelled lanthanide ion is introduced into the centre compartment of a cell with sintered glass partitions. Electrolysis is carried out for a suitable time and the activities in the anode and cathode compartments compared. The experiment is repeated at different concentrations of nitric acid.

10.3.5. *Miscellaneous applications.* There are a great many occasions when radioisotopes may be used with advantage to replace traditional methods. Examples are the measurement of solubility

(Experiments 17, 18 and 21), determination of distribution coefficients, equilibrium constants (Experiments 16, 19 and 20) and transport numbers (Experiment 22). The determination of the efficiency of a precipitation reaction such as phosphate with ammonium molybdate using ^{32}P as tracer, the separation of ions with ion exchange columns (Experiment 27) and paper chromatography of colourless substances (Experiments 24, 30, 31 and 32) are further examples.

10.4. *The synthesis of labelled compounds*

Before carrying out a particular experiment it may be necessary to prepare a labelled compound starting with a radioisotope as a simple or compound ion. The labelling of an organic compound in a particular position often demands considerable ingenuity in the manner of carrying out the synthesis. In general, it is advisable to prepare a relatively small amount of the compound with a specific activity considerably higher than is ultimately required and then mix it with inactive compound. For details of carrying out complex syntheses more advanced tests should be consulted. A few methods of carrying out relatively simple labelling will be described here.

10.4.1. *Biological synthesis of labelled sugars.* The principle is to introduce some ^{14}C as a soluble carbonate or hydrogen carbonate into a solution containing a living green plant, such as *Chlorella*, buffered to pH 5. Photosynthesis is carried out, and the sugars formed are extracted and separated by chromatography. By suitably varying the time of photosynthesis intermediate compounds in the photosynthetic cycle may be isolated (Experiment 32).

10.4.2. *Synthesis of a carboxylic acid, labelled in the carboxyl group.* The principle of the method is to carboxylate a Grignard reagent using CO_2 labelled with ^{14}C. The yield is high (80–90%). Labelled carboxylic acids are important in themselves, but they are highly valuable as reaction intermediates and this process is often the preliminary to a complex synthesis (Experiment 28).

10.4.3. *Preparation of labelled cyanide.* One process involves heating potassium metal, ammonium chloride and labelled barium carbonate in a sealed tube. Another process is to reduce labelled potassium carbonate by heating with zinc dust in a stream of nitrogen and hydrogen. A more recent method is a direct exchange between labelled carbonate and potassium cyanide. This has the advantage of simplicity.

10.4.4. *Preparation of labelled methanol.* The best method is the reduction of CO_2 obtained from labelled carbonate using lithium tetrahydroaluminate (lithium aluminium hydride), in a solvent such as tetrahydrofuranol. A small amount of methanoic (formic) acid is produced as a by-product, but the yield of methanol is high. Methanol is a valuable synthetic intermediate.

10.4.5. *Thermal exchange.* This is a modification of 10.3.1 (*c*), p. 142, in which the radioisotope is in high specific activity. An example is the preparation of labelled iodobenzene. A small amount of iodobenzene, e.g. 2–5 cm^3, containing about 10^{-4} molar fraction of free iodine, is shaken with carrier-free $^{131}I^-$ in dilute sulphuric acid. The iodide exchanges rapidly with the free iodine. The iodobenzene, now containing labelled iodine, is separated from the aqueous solution and heated under reflux at the boiling point for about 45 min. Almost complete exchange occurs. The whole experiment must be conducted in a glove box or efficient fume cupboard, and should be done in a yellow light in the absence of daylight, as organic iodides undergo photodissociation at wavelengths shorter than 600 nm.

10.4.6. *Halogen exchange using aluminium halides.* Aluminium halides exchange rapidly with gaseous halogen and they also exchange readily with organically combined halogen. This offers a convenient path to label organic halides. The rate of exchange is much more rapid with alkyl than with aryl halides.

Questions

10.1. Calculate the quantity in μCi of (*a*) ^{131}I, (*b*) ^{110m}Ag required to determine the solubility of silver iodide in water at 298 K to an accuracy of 1 % using (i) a liquid G-M tube of capacity 10 cm^3 with an efficiency of 1 % for ^{131}I and 0·5 % for ^{110m}Ag, and (ii) a scintillation counter of capacity 3 cm^3 and efficiency 30 % for both isotopes. Assume that 1 mg of silver iodide would be required to prepare the saturated solution and that a maximum of 1 h would be available for counting. A saturated solution of silver iodide is about 10^{-8} M at 298 K. Comment on the hazard and consider which isotope should be used. (See table 12.2).

10.2. Small amounts of the radioactive halogen astatine (At) may be volatilized from an irradiated bismuth target and collected in nitric acid to form a solution S. The following scheme describes experiments carried out on S. The sign * indicates that astatine activity is found in the product shown, the sign ○ that no such activity is found.

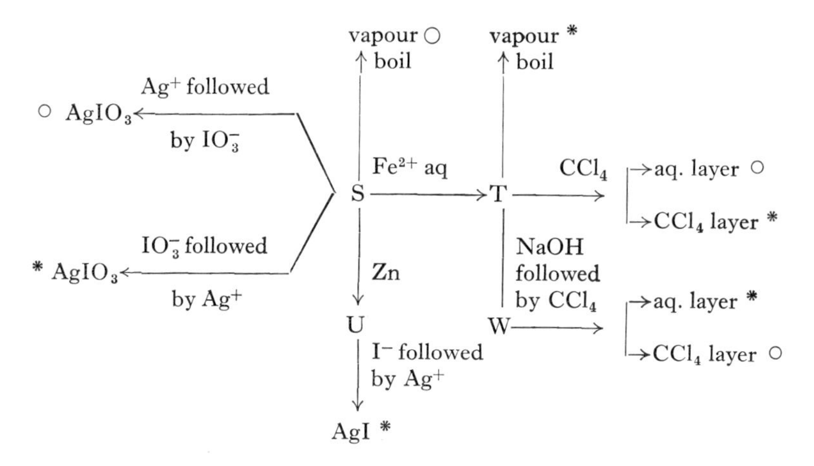

State, with reasons, what oxidation states of At are present in solutions or compounds S to W. (Oxford schol.)

10.3. The following measurements were taken in the course of an analysis of a specimen of 'Dural' for manganese, by radioactivation using a 0·5 Ci Am-Be neutron source. (Refer to Experiment 14 (*b*) for details.) Mass of $MnSO_4 . 4H_2O = 10$ g, mass of 'Dural' $= 9{\cdot}63$ g. Time of irradiation of both specimens $= 7{\cdot}84 \times 10^4$ s. Neutron flux $= 10^3$ n cm^{-2} s^{-1}. Time of decay before counting: $MnSO_4 . 4H_2O = 1{\cdot}5 \times 10^3$ s, 'Dural' $= 4{\cdot}80 \times 10^2$ s. Count of $MnSO_4 . 4H_2O = 4{\cdot}274 \times 10^3/1000$ s, count of 'Dural' $= 3{\cdot}06 \times 10^2/1000$ s, background $= 54/1000$ s. σ_a for ^{56}Mn $= 13{\cdot}3$ b (10^{-24} cm^2), ^{56}Mn ($t_{1/2} = 2{\cdot}58$ h).

Determine the percentage of manganese in the specimen quoting the accuracy of the measurement.

11. Some miscellaneous applications

11.1. *Hot atom chemistry—the Szilard–Chalmers reaction*

11.1.1. *Origin.* This reaction was first carried out in 1936 by the men whose names it bears. They bombarded iodoethane with neutrons and obtained some free radioactive iodine as ^{128}I. The reaction is of considerable interest to chemists as it involves the production of a new chemical species by a nuclear reaction. This experiment was no chance occurrence, but was carefully planned by L. Szilard [40].

Szilard was one of the most brilliant physicists of his time and also one of the laziest. A Hungarian by birth, he never had any fixed appointment, but he had small private means. These enabled him to travel round the world, and invariably to arrive when any exciting experiment was in progress. Naturally enough, he turned up at Los Alamos during the war when Fermi and his team were working on the plutonium project. Szilard asked Fermi if he could join the team, and Fermi willingly accepted him. All the members of the team were at that time working hard on constructing the pile, and Fermi noticed that Szilard was not doing any of the hard work, so he told him fairly sharply that all the members of his team were expected to do their share of the heavy work. Szilard worked for a couple of days, but then said to Fermi: ' I will pay someone to do my share of the work. I think I should be better employed sitting in a chair thinking '.

Returning to the origin of the Szilard–Chalmers reaction, Szilard duly arrived in Rome in 1934 when Fermi was doing the first experiments on bombarding elements with neutrons to see if they could be made radioactive. Szilard suggested to Fermi that he should use a thin foil of the element and place another thin foil behind, so that when the element was bombarded with neutrons the active atoms might be knocked out on to the second foil and thereby collected. The experiment was duly carried out and proved successful. It was a short step from here to try the reaction in the liquid state, and this is almost certainly how the famous reaction evolved.

The first experiment was carried out in the Medical School of St Bartholomew's Hospital, London. Chalmers was a physicist in the Medical School and had constructed β-counting equipment. He had access to the neutron sources available at the Hospital. A chemist was needed to complete the team and E. Glueckauf, who was working at the Imperial College of Science, was 'borrowed' by Szilard to do the necessary chemical separation. The iodoethane was extracted, after neutron bombardment, with sodium sulphite containing a small amount of potassium iodide. The iodide was precipitated as silver iodide and counted with an end window counter. A half-life of 25 min was recorded. (Experiment 15.)

Since the original experiment this reaction has been carried out with a great many different substances, and in the solid, liquid and gaseous states. The element concerned must be covalently bound in a molecule. When the neutron enters the nucleus an isotope in an excited state is produced, which breaks the bond uniting it to the rest of the molecule. For this reason this branch of chemistry is known as 'hot atom' chemistry.

Good examples of the reaction are the irradiation of permanganate to produce ^{56}Mn, organic bromides or inorganic bromates yielding ^{80}Br and ^{82}Br, organic chlorides or inorganic chlorates yielding ^{36}Cl and ^{38}Cl, and many other similar cases.

11.1.2. *Nature of the reaction.* The reaction between iodoethane and the neutron can be formulated as follows:

$$C_2H_5{}^{127}I + {}^1n \rightarrow C_2H_5{}^{128}I + \gamma \rightarrow C_2H_5{}^{\cdot} + {}^{128}I.$$

Bond rupture is almost certainly 100% effective (p. 152), but it was observed very early on in the course of this work that a certain proportion of the ^{128}I remained in organic combination. This proportion is known as the *retention* though the term is somewhat misleading and it should really be called *recapture*.

The percentage retention is extremely sensitive to traces of impurities, and to obtain reproducible results a small amount of iodine must be added to the iodoethane. This iodine is known as a scavenger. The percentage retention decreases with increasing concentration of scavenger. The form of the graph of retention in iodoethane against molar fraction of scavenger is shown in Figure 11.1. It will be noted that the retention falls steeply at first, but eventually becomes linear, and can be extrapolated to a scavenger molar fraction of 1, when the retention of active iodine in the organic phase becomes 0.

The first attempt to explain the retention of active iodine was due to W. F. Libby [41]. He proposed that the active iodine atom severed the bond uniting it to the rest of the molecule by recoil, as a result of the emission of γ-rays, much as a gun recoils on firing a shell. The 'hot' atom cannoned about in the liquid losing energy in the process. Most of the energy would be lost when in collision with another iodine atom, which would result in bond rupture, producing a free ethyl radical. Eventually the ^{128}I would become thermalized and would then be recaptured by an organic free radical and re-enter organic combination, or react with a molecule of scavenger iodine and enter inorganic combination. This explanation was known as the 'billiard-ball' hypothesis.

In terms of recoil the maximum energy of the hot atom can be calculated from the known maximum energy of the γ-rays emitted. The gamma spectrum of the (n, γ) reaction is complex (Figure 11.2);

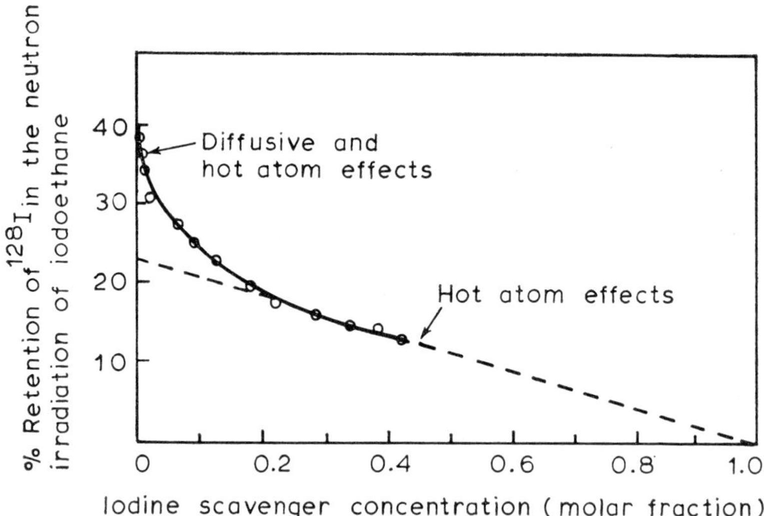

Figure 11.1. Retention of active iodine against scavenger concentration during the neutron irradiation of iodoethane. During neutron irradiation of organic halides bond rupture between the activated halogen atom and the organic radical occurs. Some active halogen atoms are recaptured, or retained in organic combination. To obtain reproducible results free halogen must be present, and this is known as a 'scavenger' because it combines with impurities. The graph shows how diffusive recombination by thermalized atoms, which are highly dependent on scavenger concentration, may be distinguished from 'hot' atom effects, which are much less dependent on scavenger concentration (see Experiment 15). (After Professor P. F. D. Shaw).

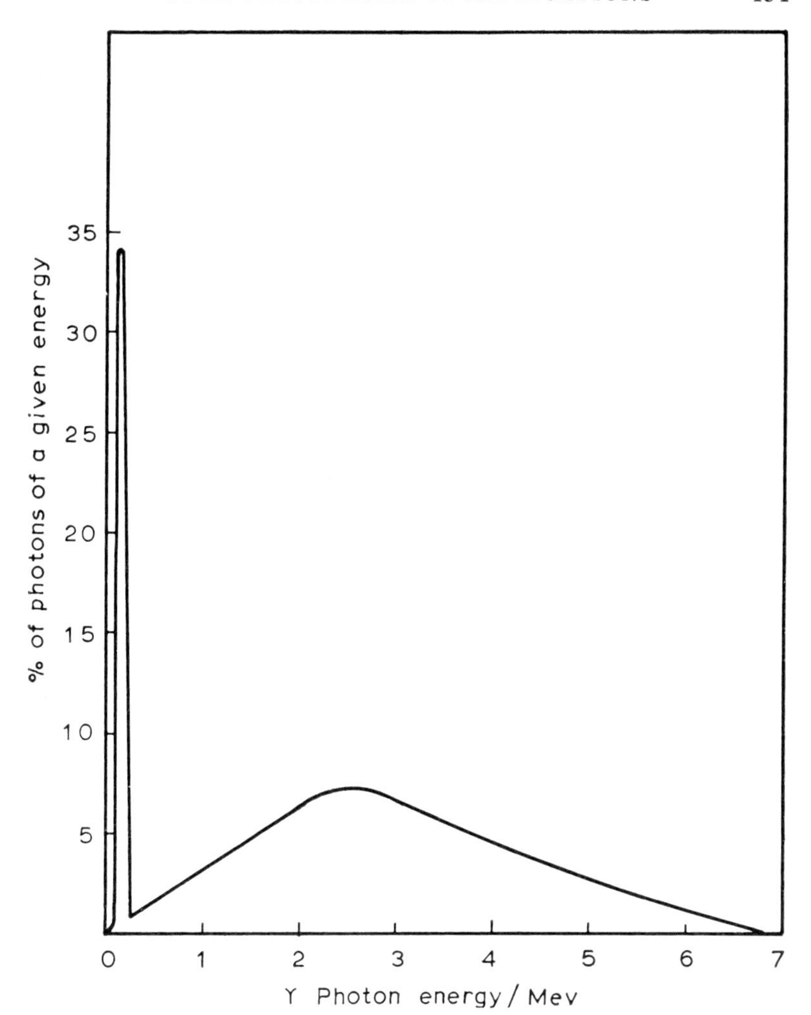

Figure 11.2. The γ-spectrum of the (n,γ) reaction during neutron irradiation of iodine. Calculation shows that 29% of the γ-ray energy of the peak at 0·135 MeV is internally converted by ejection of k-shell electrons, leading to multicharged I^{n+} which will cause charging and disruption of the surrounding molecules. (The graph was constructed by the author with considerable assistance from Professor G. N. Walton, from data obtained from references [43–45].)

the maximum energy is approximately 6·7 MeV. The calculation is performed as follows:

From the principle of conservation of momentum:

$$(mv)_{\gamma} = (mv)_{\mathrm{I}}$$

Hence:

$$(m^2v^2)_\gamma = (m^2v^2)_\mathrm{I}$$

Now:

$$(m^2v^2)_\gamma = 2m_\mathrm{I}E_\mathrm{I} \tag{11.1}$$

where E_I is the kinetic energy of the recoiling I atom.
Now:

$$v_\gamma = c$$

where c is the velocity of light,
and:

$$m_\gamma c^2 = E_\gamma$$

Hence:

$$(m^2v^2)_\gamma = (m^2c^2)_\gamma = E_\gamma{}^2/c^2.$$

So that, from Equation (11.1),

$$E_\mathrm{I} = E_\gamma{}^2/2m_\mathrm{I}c^2. \tag{11.2}$$

From Equation (11.2), the maximum energy of the recoiling I atom would be 190 eV (see Question (11.1)). The strength of the C–I bond is about 2 eV (200 kJ mol^{-1}). The hot atom, and the 'temperature' of such an atom in terms of kinetic energy would be 10^6 K, will rapidly lose energy by collision with surrounding iodine atoms. After an average of 8 such collisions, which would produce about 8 free radicals, the atom will be effectively thermalized. This process of recoil would therefore produce a maximum of 8 free radicals per event. However, experiments using scavenger iodine labelled with ^{131}I dissolved in iodobenzene have shown conclusively that at least 1000 free radicals are produced per event (published and unpublished work by the author). This indicates that there must be some process other than recoil to account for the above observation [42].

Figure 11.2 shows a low energy γ-photon peak at 0·135 MeV. 34% of the (n, γ) events produce photons of this energy. Calculation shows that 29% of this energy is internally converted by ejection of K shell electrons. This would result in the production of multi-charged positive ions with consequent charging and disruption of the surrounding molecules and the production of large numbers of free radicals. That this, in fact, does happen has been shown by gas chromatography of iodobenzene after neutron irradiation in the presence of labelled scavenger iodine. Labelled alkyl radicals have been detected amongst the products (unpublished work by the author).

An interesting case of the Szilard–Chalmers reaction is the (n, γ) reaction with an organic bromide. Apart from the useful tracer ^{82}Br, produced from ^{81}Br, two short lived products are formed from ^{79}Br; these are ^{80}Br and ^{80m}Br respectively. When the irradiated bromide is extracted with sulphite both ^{80}Br ($t_{1/2}$ = 18 min) and ^{80m}Br ($t_{1/2}$ = 4·4 h) are present in the extract, and the log graph of the decay is composite. If a second extraction is carried out some hours after the first only the 18 min decay will be found. Some of the ^{80}Br and ^{80m}Br have re-entered organic combination. The excited state of ^{80m}Br decays to the ground state ^{80}Br by isomeric transition (p. 96) with the emission of low energy γ-rays (0·05 and 0·04 MeV). 57% of the 0·05 MeV and 100% of the 0·04 MeV γ-rays are internally converted. Bond rupture producing ^{80}Br could not possibly be produced by recoil, and the alternative hypothesis of the disruption of the molecule by multi-charged ions would seem to be a reasonable explanation.

11.2. *Radiocarbon dating*

^{14}C is formed in the upper atmosphere by reaction of fast neutrons from cosmic rays with nitrogen:

$$^{14}_{7}N + ^{1}_{0}n \rightarrow ^{14}_{6}C + ^{1}_{1}H$$

The ^{14}C is eventually oxidized to CO_2. The proportion of ^{14}C to ^{12}C in the atmosphere remains constant as the rate of production is balanced by the rate of loss to outer space by diffusion, and by photosynthesis. In living matter the specific activity of ^{14}C is therefore constant. When matter dies the activity falls with the half-life of ^{14}C (5760 y). Assuming that the cosmic ray intensity has remained constant over the period of time considered, it is possible to date specimens containing carbon from the observed specific activity. The time-scale is limited effectively to 40 000 to 50 000 y by the lower limit of detectable ^{14}C activity. There is always the possibility of contamination by modern organic matter; 0·2% would give an activity equivalent to an age of 50 000 y.

The specific activity of ^{14}C in living matter is about 15 dis $min^{-1} g^{-1}$, so that a specimen which has been dead for 45 000 y will have a specific activity of $\frac{15\dagger}{256} \approx 0{\cdot}06$ dis $min^{-1} g^{-1}$. To count such a low activity very special techniques are necessary, which normally use an anticoincidence circuit, as first used by Libby [46].

† 45 000 y ≈ 8 ^{14}C half-lives; $256 = 2^8$.

11.2.1. *Proportional gas counting.* The specimen is converted to CO_2, CH_4, or C_2H_2, and introduced into a gas counter at pressures up to 3 atmospheres. A voltage of about 8 kV is applied between the central anode and the metal outer wall. In order to reduce the background the gas counter is surrounded with a ring of at least six G-M tubes connected to the anticoincidence circuit. A cosmic ray passing through the gas counter will trigger two of the G-M tubes and so will not be recorded. To reduce the background still further a thick iron or lead shield surrounds the whole apparatus, which is often housed in a room below ground with thick concrete walls. Even with this arrangement there is a low background, so that the count must be repeated under identical conditions using inert CO_2 prepared from Welsh anthracite. This second count is the background, which must be subtracted from that produced by the unknown specimen. Finally the count is compared with the count from a standard modern sample, which must be pre-1950 as nuclear tests have considerably increased the proportion of ^{14}C in the atmosphere.

11.2.2. *Liquid scintillation counting.* In this case the specimen is converted to ethyne (acetylene); this is polymerized to benzene which is then introduced into the liquid scintillator. As in the previous case, a ring of G-M tubes or liquid scintillators is arranged in anticoincidence, and heavy shielding is employed. The background is obtained by using benzene prepared from an ancient source, and the net count compared to benzene prepared from a modern source (pre-1950). A slight complication is introduced because the water used to prepare the ethyne contains traces of ^3H. The threshold in the counting circuit must be set so that the ^3H is not counted, and this reduces the efficiency of the counter for ^{14}C. With either 11.2.1 or 11.2.2 long counting times, up to 4 d are required.

^{14}C dating has been checked using dendrochronology, i.e. dating by counting tree rings from the bristlecone pine found in Arizona. This goes back for about 8000 y. The maximum divergence between the two methods is about 20%, which indicates a slight variation in the cosmic ray intensity over this period of time. There is no means at present of checking ^{14}C dating beyond 8000 y.

11.3. *Geochronology*

With the discovery of radioactivity it was recognized that radioactive decay constituted a type of clock which ran down at a rate determined by the radioactive constant, λ. Rutherford suggested a method for determining geological age based on the ratio of helium to uranium in minerals. Uranium in the course of its decay gives out eight α-particles. The α-particles rapidly capture electrons from surrounding matter and become reduced to helium which accumulates in the rock. The method gives a lower limit only to the age of the rock, because there is always the possibility of a leakage of the helium. A further complication is the presence of thorium which also produces helium [47].

With the discovery that lead was the stable end product of the uranium and thorium decay series, and that the mass number of uranium lead should be 206, while the mass number of thorium lead should be 208, a second method was suggested by Rutherford and Soddy, depending on the ratio of uranium or thorium to lead. This method was complicated by the presence of actinium lead of mass number 207 from the decay of ^{235}U, and also by the presence of non-radiogenic lead. Both these methods gave radiological ages for the earth of the order of 10^9 y, considerably greater than the early methods of Reade and Kelvin.

There are three main methods in use today for geological dating. They are:

(1) Uranium–thorium–lead

(2) Potassium–argon

(3) Rubidium–strontium

11.3.1. The uranium-thorium-lead method requires an accurate analysis of the mineral for uranium and thorium, and also the determination of the isotopic composition of the lead by mass spectrometry. If ^{204}Pb is found, this indicates the presence of non-radiogenic lead, and a correction must be made. In general the method is sound, but errors can be introduced by loss of radon, or of uranium, thorium, or other members of the decay series by leaching or diffusion.

Uranium is generally present in very low concentration. It may be determined by a number of methods. One method is by measuring the natural radioactivity using the γ-energies of ^{210}Tl or ^{214}Bi. Another is by radioactivation analysis, either to produce ^{239}Np, or fission products from ^{235}U. One such fission product is ^{140}Ba ($t_{1/2} = 12{\cdot}8$ d) which decays to ^{140}La ($t_{1/2} = 40{\cdot}2$ h). These methods have a limit of detection of about 10^{-12} g.

Thorium may be determined similarly by natural radioactivity using the γ-emissions from ^{212}Pb or ^{208}Tl. It may also be determined by using the fast neutron reaction (p. 45).

Autoradiographic techniques (see Experiment 34) may be used to detect uranium and thorium in certain parts of a rock specimen. A section of the rock is put in contact with a nuclear emulsion plate, or a liquid nuclear emulsion is poured over the rock and developed after a suitable time. The α-tracks indicate the location of the uranium or thorium.

11.3.2. The potassium-argon method is based on the decay of ^{40}K, present at 0·01% approximately in natural potassium. It is complicated by the decay scheme of ^{40}K, 89% of the events producing ^{40}Ca and 11% ^{40}Ar. Accurate dating by this method was only made possible by the exact determination of this branching ratio. The half-life of ^{40}K is $1{\cdot}26 \times 10^{9}$ y.

Potassium may be determined by analytical methods such as flame photometry using the potassium doublet 766–769 nm, neutron activation analysis in which it can be distinguished from sodium by the large difference in maximum β-energy (^{24}Na, 1·39 MeV; ^{42}K, 3·6 MeV), and by isotopic dilution analysis. Potassium can also be determined by classical analytical procedures such as the precipitation of potassium perchlorate, which is both tedious and time consuming.

Argon may be determined either by direct measurement of the total volume of the gas obtained from a sample, or by isotope dilution analysis. The second method is more sensitive than the first, and the lower limit of measurement is 10^{-7} cm^3 at STP. The method gives reliable results for non-porous rocks such as micas.

11.3.3. The rubidium-strontium method depends on the decay of ^{87}Rb to ^{87}Sr, with a half-life of 5×10^{10} y. It has the advantage over the other two methods that there is no possibility of loss of gas from the sample. In a modern sample of rubidium ^{87}Rb is present at 27·85%. In common (non-radiogenic) strontium the ratio of ^{87}Sr to ^{86}Sr is 0·702. In minerals with a high rubidium and very low strontium content the age may be determined by measuring the ratio ^{87}Rb to ^{87}Sr by mass spectrometry. In minerals with a considerable amount of natural strontium the ratio of ^{87}Sr to ^{86}Sr must also be determined.

Rubidium may be determined similarly to potassium as the perchlorate, by mass spectrometry, or by the β-activity. ^{87}Rb emits very low energy β-particles and these must be measured either with a liquid scintillation counter, or by 4π counting (so called because the whole of the emitted radiation is detected and the area of a sphere is $4\pi r^2$) with a proportional counter. The most accurate measurement of the half-life has been done by 4π counting and gives a value of $5{\cdot}25 \pm 0{\cdot}1 \times 10^{10}$ y. The rubidium-strontium method is now regarded as one of the most reliable dating methods.

Question

11.1. Given that the maximum energy of the γ-photon emitted in the ^{127}I (n, γ) ^{128}I reaction is 6·7 MeV, calculate the maximum recoil energy of the ^{128}I atom. ($1\,\text{eV} = 1{\cdot}6 \times 10^{-19}$ J, 1 mass unit $= 1{\cdot}67 \times 10^{-27}$ kg, $c = 3 \times 10^{8}\,\text{m s}^{-1}$.)

12. Radiological safety

The hazards involved in the handling of radioactive isotopes are of two kinds: (*a*) those due to external radiation and (*b*) those due to ingestion or inhalation. Students using radioactive tracers are subject to little danger from external radiation, except for spillage on the skin. They should treat the solutions with the same care with which they would treat concentrated nitric acid. The main hazard comes from ingestion or inhalation, and this hazard can be much reduced by using proper techniques (see the *Rules* on p. 167 and the instructions accompanying each experiment). For teachers and technicians who will be responsible for handling stock solutions, sealed sources and neutron sources, the hazard can be considerable and film badges and/or personal dosimeters should be worn, and if necessary finger dosimeters.

12.1. *Units of radiation*

(1) The rad. This is a measure of energy absorbed in matter, i.e. the dose, and is defined as $10^{-5}\,J\,g^{-1}$. The SI unit is the Gray (Gy), 1 J kg.

(2) The rem. This is the Rad Equivalent Man and is obtained by multiplying the dose in rads by the RBE (Relative Biological Effectiveness) for the radiation concerned. The dose in rems for β-particles, X-rays and γ-rays is essentially the same as the dose in rads. Protons, heavy ions and neutrons produce greater damage, caused in the case of the first two by increased ionization in their track. Fast neutrons produce damage from ' knock on ' protons, and thermal neutrons may produce (n, γ) effects in tissue. Table 12.1 gives the RBE for different types of radiation.

(3) MeV Curie. This is a useful practical unit to determine the rate of emission of energy from a source. 1 MeV Curie = $5{\cdot}92 \times 10^{-3}\,J\,s^{-1}$, since 1 eV = $1{\cdot}602 \times 10^{-19}$ J and 1 Ci = $3{\cdot}7 \times 10^{10}$ dis s^{-1}.

TABLE 12.1.

Type of radiation	RBE
X-rays and γ-rays	1
β-particles	1
α-particles	10
Protons	10
Heavy nuclei	20
Fast neutrons	10
Thermal neutrons	2·5

12.2. *Dose rates*

The external dose received from a source in rems depends on the strength of the source, the energy and type of radiation emitted, the time of exposure and the distance from the source. Maximum permissible levels (m.p.l.) both for radiological workers and for the general public have been fixed by international agreement. For radiological workers the dose rate to the whole body should not exceed 2·5 millirem h^{-1}, or 100 millirem per 40 hour week. There is an additional provision that the total accumulated dose to the whole body should not exceed $5(N-18)$ rem, where N is the age in years. This applies to γ-radiation.

The m.p.l. for β-radiation is more difficult to define as the β-particles are all absorbed in the surface layers. The value has been fixed at 500 millirem per week at the surface of the body. There is a relaxation in regard to extremities such as the hands, feet, and forearms. The m.p.l. is at least five times that for the whole body, i.e. about 60 millirem h^{-1} for β-radiation and 12·5 millirem h^{-1} for γ-radiation.

For members of the general public, which includes students over 18, the rate is one tenth of that for an occupational worker. This is to guard against possible genetic effects to the general population. The m.p.l. is therefore 10 millirem per week or 500 millirem per year. For students between the ages of 16 and 18, the Department of Education and Science in Britain has fixed the m.p.l. as 50 millirem per year in addition to the normal background, which is also 50 millirem per year.

12.3. *Calculation of dose rate*

This is relatively simple in the case of γ-rays, as the inverse square law applies effectively in air. The dose absorbed in tissue varies with the energy of the radiation (see Figure 12.1). For β-rays the dose rate is not so easy to calculate as absorption in air occurs and approximations have to be made. When using β-active solutions the β-particles are largely absorbed in the solvent and the remainder in the glass walls of the vessel. An ordinary glass beaker will stop completely β-particles of 1 MeV energy or less, and will reduce by a half β-particles of 3·5 MeV energy. β-particles of 1 MeV energy will deposit half their energy in 25 cm of air. Simplified formulae have been devised for calculating β and γ dose rates. For β dose rates, in the absence of the necessary tables, it is advisable to err on the side of caution and assume no absorption in air. For γ dose rates a convenient formula is:

$$6\,CE\ \mathrm{rad\,h^{-1}\ at\ 25\ cm} \tag{12.1}$$

where C stands for the strength of the source in curies and E for the energy of the radiation in MeV.

Example 1. What is the dose rate to the hands in millirem s^{-1} at 10 cm from 1 mCi of ^{137}Cs contained in a glass phial as normally supplied?

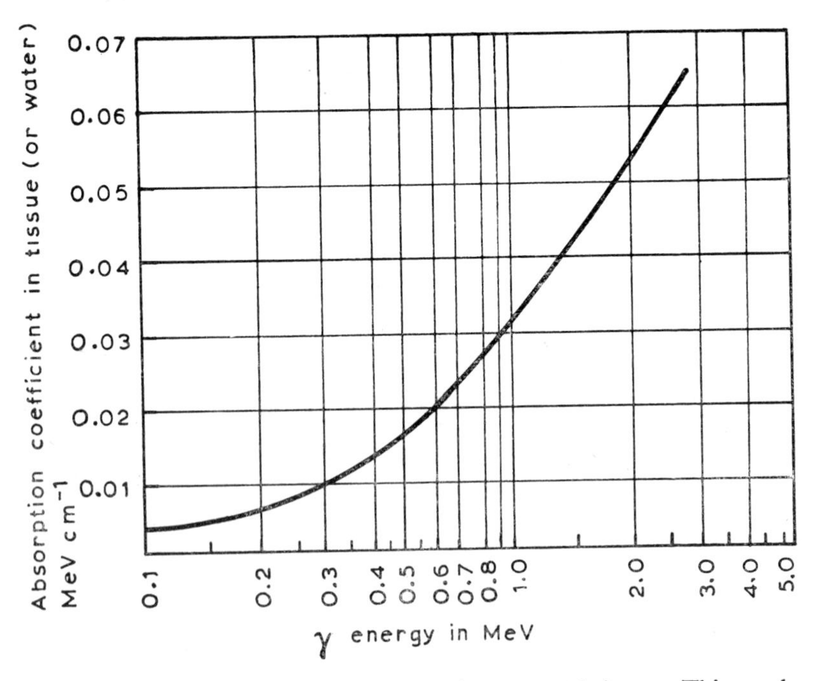

Figure 12.1. Energy absorbed by γ-rays in water and tissue. This graph may be used to determine dose rates from γ-emitters up to 3 MeV.

The β-particles from ^{137}Cs (approximate mean energy 0·22 MeV), are all absorbed in the glass. The dose rate from the 0·662 MeV γ-rays from the daughter ^{137m}Ba is calculated as follows:

the energy dissipated by the source is:

$$0{\cdot}662 \times 10^{-3} \times 5{\cdot}92 \times 10^{-3}\ \mathrm{J\ s^{-1}},$$

at 10 cm the energy flux is:

$$\frac{0{\cdot}662 \times 10^{-3} \times 5{\cdot}92 \times 10^{-3}}{4\pi \times 10^{2}}\ \mathrm{J\ s^{-1}\ cm^{-2}}$$

and the dose rate to the hands is:

$$\frac{0{\cdot}662 \times 10^{-3} \times 5{\cdot}92 \times 10^{-3} \times 0{\cdot}022 \times 10^{8}}{4\pi \times 10^{2}}$$
$$= 6{\cdot}86 \times 10^{-3}\ \text{millirem s}^{-1}$$

since 10^8 millirem = 1 J g^{-1} and 0·022 is the absorption coefficient for 0·662 MeV γ-rays in tissue expressed in MeV cm^{-1}, and the density of tissue is virtually the same as water.

Or using the approximate formula for γ-rays ($6CE$ rad h^{-1} at 25 cm):

$$\frac{6 \times 10^{-3} \times 0{\cdot}662 \times 25^{2} \times 10^{3}}{3{\cdot}6 \times 10^{3} \times 10^{2}}$$
$$= 6{\cdot}90 \times 10^{-3}\ \text{millirem s}^{-1}.$$

A radiological worker could work at this distance for a maximum of 30 min before receiving the m.p.l. to the hands of 12·5 millirem h^{-1}, or a student for about 3 min. The bottle should obviously be shielded with a lead pot. 0·5 cm of lead would reduce the radiation level by half.

If the bottle were handled, however, the dose would be about 500 times the above figure, and the m.p.l. for the radiological worker would be exceeded in 4 seconds. This shows the importance of handling all such bottles with tongs. In the case of Experiment 5 (half-life of ^{137m}Ba) the student uses 0·2 μCi; he can safely handle a test-tube, containing this quantity, for 30 minutes.

Example 2. What would be the dose rate if 0·2 μCi of ^{137}Cs were spilled on the skin? The mean energy of the β-particles from ^{137}Cs is 0·22 MeV.

(*a*) dose from γ-rays:

$$\tfrac{1}{2}(0{\cdot}662 \times 0{\cdot}2 \times 10^{-6} \times 5{\cdot}92 \times 10^{-3} \times 0{\cdot}022 \times 10^{8})$$
$$= 8{\cdot}62 \times 10^{-4}\,\text{millirem s}^{-1}$$

(*b*) dose from β-rays:

$$\tfrac{1}{2}(0{\cdot}22 \times 0{\cdot}2 \times 10^{-6} \times 5{\cdot}92 \times 10^{-3} \times 10^{8})$$
$$= 1{\cdot}30 \times 10^{-2}\,\text{millirem s}^{-1}$$

which gives a total dose rate of $1{\cdot}38 \times 10^{-2}$ millirem s^{-1}. A student would, therefore, receive his m.p.l. for 1 h in about 7 min. This shows the importance of careful handling and immediate washing of the skin in the event of a spill. However, the chance of spilling the entire contents of the test-tube on the hands is remote. The reader is referred to Question 12.2 which deals with dose rates from different commonly used radioisotopes.

12.4. *Neutron sources*

Precautions must be taken in handling these sources owing to the hazard from high energy neutrons and the γ-radiation accompanying the neutron emission. Am–Be sources give out neutrons with energies up to 10 MeV. They should be kept in a large block of paraffin wax or polythene, about 25 cm radius. If the source is withdrawn for any purpose it must be remembered that a flux of 20 high energy (1 MeV) neutrons cm^{-2} s^{-1} will produce a dose rate of 2·5 millirem h^{-1}. This flux would be produced by a 1 Ci source at 90 cm. The corresponding dose from thermal neutrons would be produced at 45 cm. The dose rate from neutron sources when in a block of moderator should be checked at various distances with a neutron monitor.

As far as external radiation from radioactive sources is concerned the two most easily controllable factors are time and distance; i.e. keep as far away, and handle for as short a time, as is consistent with safe working, and NEVER handle sources with the fingers.

12.5. *Hazards from ingestion or inhalation*

Internal doses are potentially much more hazardous than those received externally. The hazard obviously depends on the quantity ingested and on the type of radiation, but it also depends on whether the isotope is taken up by some specific organ of the body, or whether it is generally distributed, and whether or not it is rapidly excreted. Inhalation of volatile radioactive material is particularly hazardous. Owing to the large volume of air inhaled, about 15 dm^3 min^{-1}, the

MPC_a (maximum permissible concentration in air) is fixed at a very low value. ^{131}I for example has an MPC_a of 6×10^{-9} μCi cm^{-3}. ^{131}I is an isotope of medium toxicity. The hazard caused by the inhalation of radon in the uranium mines of Saxony, where the radon concentration is 3×10^{-6} μCi cm^{-3}, resulted in the death of 50% of the miners from lung cancer.

Various criteria for safety have been laid down internationally such as the MPC_w (maximum permissible concentration in water), MPBB (body burden) and MPAI (annual intake). The first three (including the MPC_w) will be found in ICRP (International Council of Radiological Protection) publication 2. The MPAI is not so readily accessible. As it is a useful quantity to consider in tracer work the value for some tracers in common use is given in Table 12.2. AM 65, published by the Department of Education and Science, states that the maximum amount of any radioisotope used by students under the age of 18 per single experiment must not exceed one tenth of the MPAI. In most experiments only a small fraction of this quantity is required. This is certainly true for the great majority of the experiments described in this book.

TABLE 12.2. *Maximum Permissible Annual Intake (MPAI).*
(The values are given in μCi for soluble compounds ingested.)

Nuclide	MPAI
^{3}H	$2{\cdot}4 \times 10^4$
^{14}C	$6{\cdot}4 \times 10^3$
^{24}Na	$1{\cdot}6 \times 10^3$
^{32}P	$1{\cdot}6 \times 10^2$
^{35}S	$4{\cdot}8 \times 10^2$
^{36}Cl	$6{\cdot}4 \times 10^2$
^{45}Ca	$7{\cdot}2 \times 10^1$
^{56}Mn	8×10^2
^{60}Co	4×10^2
^{65}Zn	8×10^2
^{82}Br	$2{\cdot}4 \times 10^3$
^{90}Sr	$3{\cdot}2$
^{110m}Ag	$2{\cdot}4 \times 10^2$
^{131}I	$1{\cdot}6 \times 10^1$
^{137}Cs	$1{\cdot}6 \times 10^2$
^{198}Au	4×10^2

The hazard from ingestion and inhalation can be reduced to a minimum by the use of proper techniques. For example, all work using volatile radioactive material should be conducted under an efficient fume hood, in a glove box, or in a vacuum line. Work with non-volatile materials must be conducted over a tray and all apparatus marked with special tape marked 'radioactive'. No mouth operations may be permitted. Hands must be washed and monitored before leaving the laboratory. Laboratories must be provided with at least one (β, γ) monitor. If the rules for safe working, printed on p. 167 are strictly followed the hazard should be negligible.

For further study the reader is referred to publications such as *Radioisotope Laboratory Techniques* by Faires and Parks, *The Code of Practice*, HMSO and *AM* 65, DES.

12.6. *Handling techniques*

The majority of the experiments described in this book do not require the use of gloves on the part of the student. Gloves constitute a hazard with inexperienced students particularly when wet and may lead to spills. However, students should have practice in the use of gloves and gloves have been recommended for certain experiments. As instruments and reagent bottles must not be handled with possibly contaminated gloves, either the gloves must be removed or tissues used. This is time-consuming. If gloves are required it is often preferable to use only one for handling the vessel containing the radioactive isotope. This leaves the other hand free to operate instruments and handle reagent bottles. Before rubber gloves are put on the hands should be rubbed with French chalk.

When removing gloves the first glove should be removed with the other gloved hand and the second glove removed by inserting the fingers of the free hand inside the second glove. The outside of a glove must NOT be touched with a bare hand. Polythene disposable gloves are preferable to rubber for student use. They are cheap and are easily removed from the hand with tongs.

If gloves are slightly contaminated they should be immersed in 'Decon' or other decontaminating agent. Seriously contaminated gloves must be discarded as solid waste.

Teachers and technicians will certainly need gloves for dispensing radioactive solutions. They will also need long-handled tongs for handling neutron or γ-sources. The proprietary makes are expensive, but for many purposes the long-handled tongs called 'Long Arm Reachers', which are used by shopfitters, are quite suitable and very much cheaper. They are made in two sizes, 1 m and 1·5 m. Both sizes are useful.

Highly active sources must be handled behind a lead wall composed of lead bricks. These are expensive to buy, but they can easily be cast from scrap lead in a sand mould. Transparent bricks can be made from 3 mm Perspex and filled with a saturated solution of lead nitrate or zinc chloride. With the aid of these transparent bricks operations conducted behind the wall can be observed.

12.7. *Decontamination*

Apparatus is most easily decontaminated by soaking in a 2–5% solution of Decon or other suitable complexing and wetting agent. Some nuclides, notably ^{32}P, are more difficult to remove than others, and prolonged soaking or warming may be necessary. Iron(III) chloride is a very useful scavenger, particularly for removing uranium and its daughter products.

Benches should always be covered with 'Benchcote' or other suitable absorbent backed paper, and floors with linoleum, well polished. In the event of a serious spill a portion of the floor covering may be removed and destroyed.

12.8. *Disposal of waste*

Before radioactive waste, either solid or liquid, may be disposed of to the refuse or drainage system, the prior consent of the Local Authority must be obtained. Small quantities, of the order of a few μCi, may be poured down the drain. A drain should be chosen which has direct access to the drainage system and the sink should be marked 'For Radioactive Waste'. The discharge should be well flushed with water. The quantity of radioactive waste which can be safely disposed of through the sewage system depends on the daily volume of effluent. Suppose this were 10^4 gallons per day. If 1 mCi of activity were poured down the drain this would produce a concentration of $2{\cdot}5 \times 10^{-5}\,\mu\mathrm{Ci\,cm^{-3}}$ in the sewer. This is below the $\mathrm{MPC_w}$ for all the radionuclides mentioned in Table 12.2, except ^{90}Sr (4×10^{-6}). The concentration would be reduced much further at the sewage works.

Solid waste is more difficult to deal with. Slightly contaminated tissues must be placed in a specially marked bin and subsequently mixed with solid refuse. There is a general condition that the maximum amount of solid radioactive waste disposed of through the general refuse system must not exceed 10 μCi per week, and not more than 1 μCi for any one article. Short lived products should be stored until they have decayed sufficiently to permit of disposal. If it is necessary to dispose of larger quantities than those already mentioned special permission must be obtained, and the Radiological Protection Service in Britain should be consulted.

Questions

12.1. A radiological worker ingests into his lung a particle of $^{239}PuO_2$, of diameter 10^{-6} m with an activity of 0·28 pCi. Calculate the dose he receives (*a*) to his whole lung, (*b*) to the local tissue, (*c*) to one human cell in rems y^{-1}. [Mean energy of α-particles emitted by $^{239}Pu = 5{\cdot}14$ MeV. Range of an α-particle of this energy in body tissue approximately $3{\cdot}5 \times 10^{-5}$ m. Density of body tissue may be taken as equal to water. The RBE (p. 159) of α-particles is 10. 1 Rad $= 10^{-2}$ J kg^{-1}. Volume of a human cell $\approx 4 \times 10^{-15}$ m^3.] Comment on the answers.

12.2. What is the dose rate to the fingers in millirem s^{-1} (*a*) at 10 cm, (*b*) at 1 mm (contact), from 1 mCi of each of the following nuclides: (i) ^{32}P, (ii) ^{60}Co, (iii) ^{14}C? Assume a small source.

Calculate the time taken to receive the maximum permissible hourly dose (60 millirem for β-particles and 12·5 millirem for γ-rays for a radiological worker).

Assume the following data:

Energy of γ-rays emitted by ^{60}Co, 1·17 and 1·33 MeV, both immediately following β-emission.

Energy deposited in tissue per cm of path by ^{60}Co γ-rays $= 0{\cdot}032$ MeV.

Energy loss in air for ^{32}P β-rays (mean energy 0·57 MeV) negligible over 10 cm.

Energy loss in air for ^{60}Co β-rays (mean energy 0·1 MeV) $= 2{\cdot}5 \times 10^{-3}$ MeV cm^{-1}.

Energy loss in air for ^{14}C β-rays (mean energy 0·052 MeV) $= 7 \times 10^{-3}$ MeV cm^{-1}.

Comment on the answers.

RULES FOR WORKING IN THE RADIOCHEMICAL LABORATORY

1. No drinking, eating or smoking is permitted in the laboratory.
2. No mouth operations may be performed. All pipetting must be done with syringes. Labels must NOT be moistened by mouth.
3. Laboratory coats must be worn at all times.
4. Taps, switches and monitors must NOT be handled with gloves. If gloves are worn, instruments must be handled with a tissue.
5. All work with open sources, i.e. liquids or solids, must be conducted over a tray.
6. Solid waste must be placed in the active bin.
7. Liquid waste must be poured into the ' active residues ' bottle.
8. Gloves, coats *and hands* must always be monitored before leaving the laboratory.
9. Glassware must be monitored and decontaminated after an experiment.
10. In the event of a liquid spill:
    (*a*) Drop a handful of paper tissues on the site of the spill.
    (*b*) Put on gloves;
    (*c*) Mop up the spill with paper tissues;
    (*d*) Mark the spot;
    (*e*) Report the spill immediately;
    (*f*) Monitor the surface when dry.

Experiments with radioactive material

The experiments described in the following pages have all been tested in the author's laboratory. A number of them are, to the best of his knowledge, original. Experiments 23 and 30 are reproduced by kind permission of Professor Odell of Auckland University.

Experiments 1–13 are designed to illustrate some of the more important aspects of radioactivity. The remainder are examples of the use of isotopes to perform experiments which either cannot be done without their aid, or which can be carried out more simply and more quickly using radioisotopes, than by conventional methods.

For the majority of the experiments G–M tubes can be used, but it must be stressed that in many cases a well-type scintillation counter has enormous advantages over the G–M tube, owing to its greater sensitivity and ease of operation. This increases the safety margin as less activity is required. A neutron source must be used for five of the experiments and is desirable for nine. Neutron sources are a heavy initial outlay, but they save trouble and expense in purchasing some of the short-lived isotopes, and give increased safety by using ^{128}I instead of ^{131}I.

Apparatus

Apart from the normal apparatus of a chemical laboratory the following special items will be required:

Essential items

(1) Scaler/timer with quenching unit.

(2) G–M tube with thin end window.

(3) Tube holder and castle (a tube holder can be made in the workshop from perspex, fitted with ledges to take sources and absorbers at different distances from the tube, and the lead castle can be cast from scrap lead (see Appendix 7)).

(4) Liquid counting G–M tube of the M6H type with castle (again this castle may be cast in the workshop, Appendix 7).

(5) Monitor (simple β, γ type).

(6) Set of aluminium absorbers, and set of aluminium or stainless steel planchets.

(7) 1, 2, 5 and 10 cm^3 pipettes fitted with syringes (ordinary graduated pipettes are quite satisfactory if fitted with plastic syringes attached with a short length of rubber tubing).

(8) A supply of X-ray film (12 in × 10 in for Experiment 32).

(9) 10 and 50 μlitre syringes for dispensing isotopes from stock solutions.

(10) Trays lined with Benchcote, box of tissues, rubber or polythene gloves, bin marked ‘ solid radioactive waste ’.

With the above items 24 of the 35 experiments can be carried out.

Desirable items

(11) Well-type scintillation counter (for Experiments 5, 13 and 20, and desirable for 17, 18, 19, 21, 22 and 25).

(12) High voltage and pulse height analyser (could form part of the scaler/timer).

(13) γ-spectrometer (Experiments 9, 14 and 27).

(14) Liquid scintillation counter (Experiments 12 and 29).

(15) Neutron source (Am–Be suggested of strength 0·5–1 Ci)—for Experiments 8, 11 (*c*), 13, 14 (*b*) and 27 (*b*) and desirable for 15, 16, 25 and 26.

The following isotopes will be required, apart from uranium and thorium compounds: ^{137}Cs, ^{60}Co (sealed source), ^{90}Sr, ^{32}P, ^{14}C, ^{3}H, ^{110m}Ag, ^{131}I, ^{24}Na, ^{42}K.

Notes on the above-mentioned isotopes:

^{137}Cs, required for Experiments 5 and 9, very useful for checking performance of a scintillation counter and calibrating a γ-spectrometer. 100 μCi of stock solution would be sufficient.

^{60}Co, sealed source of 1 μCi required for Experiment 9.

^{90}Sr, standardized solution of 1 μCi in 5 cm^3 of 0·1 M HNO_3 for Experiment 6.

^{32}P, as sodium phosphate obtainable as a stock solution, for Experiments 10, 23 and 33. 100 μCi would be sufficient.

^{14}C, as sodium carbonate or hydrogencarbonate for Experiments 12, 28, 29, 30, 31 and 32. 1 mCi would be sufficient.

^{14}C, standardized 2 μCi in hexadecane for Experiment 12.

^{3}H, standardized 7 μCi in hexadecane for Experiment 12.

^{110m}Ag, as stock solution for Experiments 19, 21 and 22. 100 μCi would be sufficient.

^{131}I, as sodium iodide solution (1·5 μCi)—only required for Experiment 20 but if no neutron source available, also required for Experiments 16 and 26.

^{24}Na and ^{42}K, obtained as chlorides by reactor irradiation, 50 μCi sufficient for Experiment 24.

All the experiments can be carried out with μCi quantities well within the limits laid down in AM 65 by the DES. However, for those not working under the Code of Practice it will be necessary to obtain permission to hold stocks of more than half of the MPAI for any particular isotope. Special permission must be obtained to hold any quantity of ^{90}Sr.

Precautions which should be taken when using liquid counting G–M tubes. These tubes must NEVER be filled in the castle, to avoid contamination, and also to avoid the risk of short-circuiting the connections to the tube. They may be conveniently supported with a terry clip attached to a piece of dowel rod which may be held in a retort stand. Before replacing them in the castle they should be wiped with a tissue. In addition the high voltage must be switched off before the tube is withdrawn. When the tube is replaced and the high voltage switched on again a few seconds must be allowed for stability to be reached before counting commences.

All G–M tubes should be used with a quenching unit. This both lengthens the plateau and prolongs the life of the tube considerably. G-M tubes should NOT be handled with gloves.

Experiment 1. The statistics of radiochemical measurements

Theory. This experiment is designed to show the type of variation which can be expected in a series of radiochemical measurements. It has also an important function in that it can be used to check a counting assembly to see that it is working correctly.

Apparatus and materials. Scaler/timer with G–M tube and quench unit, or scintillation counter; source of long half-life such as ^{137}Cs to give 2000–5000 counts min^{-1}.

Experimental details. Arrange the source at a fixed distance from the detector. If a well-type scintillation counter is used a suitable strength for the ^{137}Cs source is 0·01 μCi, or 0·1 μCi for a G–M tube. Take 20–30 counts for 60 s or such a time that at least 2000 counts are recorded. Tabulate the results.

The following determinations should now be made:

(1) The mean count ($\bar{x}$).

(2) The individual deviations from the mean count (Δ) and the square of each deviation (Δ^2).

(3) The square root of the average of the Δ^2 terms, i.e.,

$$\sqrt{(1/n\Sigma\Delta^2)}$$

This is equal to the standard deviation (σ).

(4) The algebraic sum of the deviations $\Sigma\Delta$ which should be nearly zero.

(5) The mean deviation $\bar{\Delta}$, i.e. $\Sigma\Delta$s taking signs as positive, divided by n.

In order to check that the counter assembly is working properly carry out the following tests:

(1) Take the square root of $\bar{x}$. This should be close to the standard deviation.

(2) Divide the mean deviation $\bar{\Delta}$ by the standard deviation σ. The ratio should be nearly 0·8.

(3) Count the number of times the actual deviation (Δ) exceeds the standard deviation (σ). This should be about one-third of the number of observations.

(4) Note how many values of Δ are twice the standard deviation (σ). There should be about 1 in 20.

Precise agreement cannot be expected, but this type of test will reveal such faults as a general increase in the number of counts due to a damaged tube, or spurious counts such as occur with faults in the electronic recording system. The larger the number of observations the more precise will be the agreement.

A typical series of results is shown below:

A (counts min^{-1})	B (Δ)	C (Δ^2)
2539	−80	6400
2652	+33	1089
2612	− 7	49
2530	−89	7921
2660	+41	1681
2645	+26	676
2587	−32	1024
2574	−45	2025
2597	−22	484
2641	+22	484
2564	−55	3025
2665	+46	2116
2668	+49	2401
2677	+58	3364
2572	−47	2209
2577	−42	1764
2698	+79	6241
2720	+101	10201
2551	−68	4624
2648	+29	841

(1) Add Column A = 52 377
Divide by 20 = 2 618·9 ≃ 2619
This is the mean count ($\overline{x}$).
Column B represents the variation of the actual count from the mean count and is denoted by Δ.

(2) Add (algebraically) Column B: = −3
which is very close to zero.
Column C gives values of Δ^2.

(3) Add Column C: $\Sigma\Delta^2 = 58619$
Divide by 20, $(1/n)\Sigma\Delta^2 = 2932$
The Standard Deviation is given by $\sqrt{((1/n)\Sigma\Delta^2)} = 54{\cdot}14$
and is denoted by the symbol σ.

(4) Sum the variations of $\Delta = 971$
Mean deviation $= 971 \div 20 = 48{\cdot}6$

(5) The results of (3) and (4) give a value for the ratio

$$\frac{\text{Mean deviation}}{\text{Standard deviation}} = \frac{48{\cdot}6}{54{\cdot}14} = 0{\cdot}897$$

which is reasonably close to the value 0·797.

(6) Further investigations show that:

(*a*) The number of times the actual deviation exceeds the standard deviation $=7 \simeq \frac{1}{3}$ of the number of observations

(*b*) The actual deviation is very nearly double the standard deviation in one case of 20 readings, namely that of the 18th observation.

(*c*) The square root of the mean count. $\sqrt{2619}=51{\cdot}2$.
This is nearly equal to the standard deviation.

Experiment 2. Growth and decay

Theory. Radioactive decay processes often give rise to daughter products which themselves decay. If the rate of decay of the daughter is greater than that of the parent a state of equilibrium is reached in which the rate of growth of the daughter becomes equal to the rate of decay (Chapter 5, p. 69). If the daughter product is separated from the parent the rate of decay of the daughter can be measured and similarly the rate of growth of the daughter in the parent (Figure 1.1, p. 5).

Many such parent/daughter pairs exist but most of them decay at inconvenient rates. In this experiment use is made of a readily obtainable thorium compound and one of the daughter products is separated chemically (see Figure 1.2, p. 7).

All the daughter products will be in radioactive equilibrium with ^{232}Th provided the compound in question has not been separated for 30–40 years. In freshly separated thorium compounds ^{228}Ra and ^{228}Ac will be missing and the activity of the compound will fall for the first few years, but will gradually recover (see Experiment 17 for more detail).

When a solution of ammonia is added to a solution of a thorium salt thorium hydroxide will be precipitated along with lead hydroxide as ^{212}Pb (see also caption to Figure 1.1). The ^{224}Ra and ^{228}Ra, if present, will remain in solution. Clean separation is achieved by addition of traces of lead and barium salts as carriers. The barium can be separated as sulphate, carrying down the radium isotopes. The β-activity of the barium sulphate will initially be zero, but will increase with time and reach a maximum in about $2\frac{1}{2}$ days. The β-activity of the thorium hydroxide containing the ^{212}Pb will fall since the ^{212}Pb has been separated from its parent, and reach a minimum in about $2\frac{1}{2}$ days. Subsequently the growth curve will slowly fall since the ^{224}Ra has been separated from the parent ^{228}Th, and the decay curve will slowly rise, since ^{212}Pb will be reformed due to the gradual re-establishment of the chain from ^{228}Th.

Apparatus. Scaler/timer with quench unit, end-window G–M tube, planchets, tube holder with castle and supports for planchets, infrared lamp, small centrifuge, test pipette, set of semi-micro test-tubes, tray with tissues.

Materials. Thorium hydroxide or nitrate, 1% solutions of lead and barium nitrates, concentrated nitric acid, concentrated ammonia, dilute sulphuric acid (M).

Experimental details. Prepare a 10% solution of thorium nitrate. If the so-called hydroxide (really a hydrated oxide) is used it must be dissolved in fairly concentrated nitric acid and the solution evaporated to dryness on the water bath. Take 1 cm^3 of the 10% solution, transfer to a semi-micro test tube and add 3–4 drops of the barium and lead nitrate solutions respectively. Next add 3–4 drops of 0·880 ammonia solution, which must be free from carbonate. Stir, warm and centrifuge. Remove the centrifugate with a teat pipette, transfer to another semi-micro test tube and retain the precipitate of thorium hydroxide. Precipitate the barium in the centrifugate as sulphate by adding a few drops of dilute sulphuric acid. Centrifuge and remove most of the centrifugate, slurry the precipitate of barium sulphate, withdraw the slurry with the teat pipette and transfer to an aluminium planchet. Evaporate to dryness under the infrared lamp. Cover when dry with sellotape.

Start the scaler, measure the background, and count the precipitate of barium sulphate. Record the time. The activity should be zero.

While the count is proceeding redissolve the precipitate of thorium hydroxide in 3–4 drops of concentrated nitric acid, warm if necessary, cool and reprecipitate with 0·880 ammonia. It is necessary to use concentrated reagents to keep the total bulk small. Warm, stir and centrifuge. Withdraw the centrifugate, add water, stir and recentrifuge. Thorough washing is essential to remove adhering barium and radium ions. Slurry the precipitate withdrawn with a teat pipette, and add a little at a time to another aluminium planchet placed under the infrared lamp. Evaporate to dryness and collect the precipitate together with a small spatula. Cover with cellotape and count for about five minutes. The initial activity may be 200–300 counts min^{-1}. Record the time. Repeat counts of both specimens at convenient intervals over several days and plot the results against the time in hours.

Health hazard. There is no appreciable health hazard owing to the small quantities involved, but chemical operations should be carried out over a tray.

Experiment 3. To separate uranium 'X' from uranium

Theory. This is the classical experiment first performed about 1900 by Sir William Crookes (p. 3), in which he showed that by a single chemical operation nearly the whole of the radioactivity can be removed from a uranium compound. Uranium 'X' is in radioactive equilibrium with uranium, provided the uranium has not been treated chemically for six months. Uranium 'X' is composite, consisting of UX_1 (^{234}Th, $t_{\frac{1}{2}}=24$ d) in radioactive equilibrium with its daughter product UX_2 (^{234}Pa, $t_{\frac{1}{2}}=72$ s). The principle of the chemical separation depends on the fact that uranium ions form a complex with ammonium carbonate, whereas thorium ions do not, but form an insoluble carbonate. Owing to the very small amount of ^{234}Th present in uranium, about 1 part in 10^{10}, no precipitate is formed in the absence of a carrier. Crookes owed the success of his experiment to the presence of small quantities of impurities, notably iron(III) salts, and the fact that he measured the activity by the photographic method, i.e. by detection of β-activity only, rather than α-activity (see also p. 4).

Apparatus. Scaler/timer with quench unit, liquid counting G–M tube, suitable tube-holders with castles, small Hirsch funnel and filter flask.

Materials. Uranyl(VI) nitrate, 2% solution of iron(III) chloride, ammonium carbonate solution.

Experimental details. Dissolve about 1 g of the uranium salt in 5 cm^3 of water in a beaker and add 3–4 drops of iron(III) chloride solution. Test the activity of this solution in the liquid counter.

Return the uranium solution from the liquid counter to the beaker and add ammonium carbonate solution until the precipitate first formed almost completely redissolves, leaving a faint suspension of iron(III) carbonate. Filter off this precipitate (uranium 'X') on to a small piece of filter paper placed on the Hirsch funnel, using a water pump, and retain the filtrate. Remove the filter paper with tongs, dissolve the precipitate off the filter paper with dilute hydrochloric acid and count in the liquid counter. Measure the total volume of the filtrate and count a measured volume (5 or 10 cm^3 according to the capacity of the liquid counter). Determine the total count of the filtrate and compare this with the original count. The filtrate and uranium 'X' should be set aside, and if counted at weekly intervals growth and decay curves can be established.

Health hazard. There is no health hazard, but it is advisable to carry out the operation over a tray lined with tissues and treat the uranium as a slightly poisonous chemical.

Decontamination of the liquid counter. Uranium salts are difficult to remove, but if a solution of iron(III) chloride is left in the tube for some hours effective decontamination will result.

Experiment 4. To separate ^{208}Tl from thorium nitrate and to determine the half-life

Theory. ^{208}Tl, a decay product of the thorium series is formed from ^{212}Bi, which has a branching decay thus:

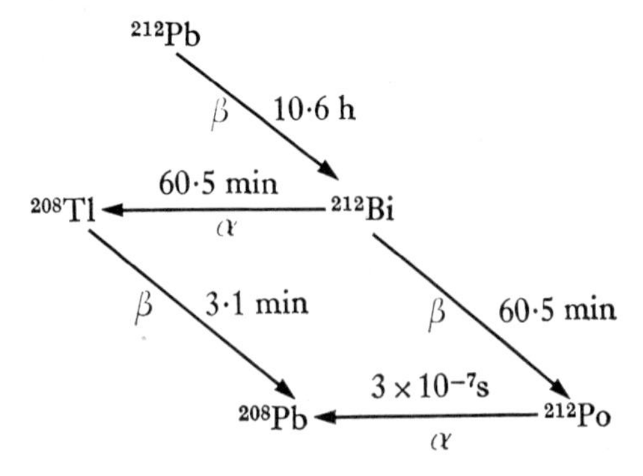

37% of the ^{212}Bi atoms decay to ^{208}Tl, which can be extracted from the thorium nitrate solution with a suitable organic solvent.

Apparatus. Scaler/timer with quench unit, liquid counting G–M tube with castle, small retort stand and clamp, 100 cm^3 beaker, 25 cm^3 measuring cylinder, 50–100 cm^3 separating funnel.

Materials. Concentrated thorium nitrate solution (40–70%), concentrated hydrochloric acid, bromine water, 2-methylethoxypropane (di-isopropyl ether).

Experimental details. Take 10 cm^3 of the thorium nitrate solution and add 10 drops of bromine water (by dropper in the fume cupboard). Add 10 cm^3 of concentrated hydrochloric acid. (N.B.—If the thorium nitrate solution is more dilute than 40% more than 10 cm^3 should be taken, depending on the concentration. An equal volume of hydrochloric acid must be used.) Pour the mixture into the separating funnel and shake for about 30 s with 11–12 cm^3 of di-isopropyl ether. The thallium is extracted as Tl(III). Rapid phase separation occurs. Run out the lower layer (aq) into the thorium residues bottle and run the upper layer (ether) into the liquid counting G–M tube which must be clamped in the retort stand, NOT in the castle. Transfer the G–M tube into the castle and count for 40 s at 1 min intervals. Continue counting for at least 12 min.

Determine the background count with the G–M tube filled with di-isopropyl ether. Subtract the background from each of the readings, and plot the log of the count rate against the time (or use semi-log paper). Measure the slope of the graph which gives λ (the disintegration constant). Determine the half-life from the relation:

$$t_{\frac{1}{2}}=0{\cdot}693/\lambda$$

Health hazard. There is no appreciable health hazard, but chemical operations must be carried out over a tray and care must be taken not to spill the concentrated solution of thorium nitrate.

Experiment 5. To separate ^{137m}Ba from ^{137}Cs and to determine the half-life

Theory. ^{137m}Ba is in radioactive equilibrium with ^{137}Cs and may be separated by precipitation as phosphate using a barium salt as carrier. The half-life of the ^{137m}Ba is 2·63 min (see Chapter 5, p. 68). 88% of the nuclei decay by isomeric transition (p. 96), with the emission of a 0·66 MeV γ-ray, to ^{137}Ba, which is stable. The remainder decay by β^- emission to ^{137}La, which has a long half-life. The rate of decay can be studied with the aid of a scintillation counter. A G–M tube could be used, but the efficiency is low, about 1%. This experiment teaches the method of achieving a clean separation of a genetically related pair.

Apparatus. Scaler/timer with (preferably) a well-type scintillation counter, or liquid counting G–M tube, with quench unit, small centrifuge for semi-micro test-tubes, syringe with hypodermic needle, teat pipette, beaker for radioactive waste, tray and tissues.

Materials. Solution of ^{137}Cs, barium chloride solution, about 2%, sodium phosphate solution about 0·2M, caesium chloride or nitrate, dilute hydrochloric acid.

Experimental details. Switch on the scaler/timer and a few moments later switch on the high voltage. Using a source of ^{137}Cs of about 0·1 μCi strength adjust the high voltage and discriminator (or gating circuit if one is fitted) to give optimum counting conditions (see Experiment 9) (if a G–M tube is to be used, set the high voltage about 50 V above the start of the plateau). Count the background for 10–20 min. While the background is being counted start the experiment.

Add about 0·5 cm^3 of the barium chloride solution to a semi-micro test-tube and dissolve a few crystals of the caesium salt in the solution. This is to act as a 'hold-back' carrier, i.e., to reduce adsorption of the ^{137}Cs on the precipitated barium phosphate. Withdraw about 0·2 μCi of ^{137}Cs from the stock bottle (2 μCi for use with a G–M tube), using the hypodermic needle, and transfer to the semi-micro test-tube. Precipitate the barium as phosphate with a few drops of sodium phosphate solution, centrifuge and discard the centrifugate to radioactive waste, i.e. put it in the marked beaker. Redissolve the precipitate with a few drops of dilute hydrochloric acid, add a few crystals of the caesium salt, and reprecipitate the phosphate by addition of sodium hydroxide solution. Again centrifuge and discard the centrifugate to waste. Repeat the whole operation. The object of this is to remove traces of ^{137}Cs which have been trapped in the crystals of the barium phosphate precipitate. Finally wash the precipitate and again centrifuge discarding the washings to radioactive waste. Wipe the outside of the semi-micro test-tube carefully and place it in the well of the counter. Take particular care not to contaminate the counter. It is advisable to wrap the tube in a tissue. If a well-type scintillation counter is not available, slurry the barium phosphate precipitate into a small beaker and place this on the top of the counter.

If G–M counting is to be used redissolve the barium phosphate in dilute hydrochloric acid, make up to the approximate volume of the G–M tube, using a measuring cylinder, and pour into the G–M tube.

These operations need not take more than about 8 min, i.e. about three half-lives. Count for 30 s at 1 min intervals, for 12 min. With a scintillation counter the initial count should be about 4000 and with a G–M tube about 200.

Plot the counting rate against the time and draw a *smooth curve* through the points. Take the initial count as the count at $t=0$. Take a series of readings from the graph of counting rate against time at 1 min intervals. Plot the log of the counting rate against the time or use semi-log paper. Provided a clean separation has been achieved a good straight line will result. From the graph determine the half-life.

Health hazard. ^{137}Cs is a medium toxicity nuclide. The quantity used, even with the G–M tube, is 1/80th of the MPAI, but all wet operations must be conducted over the tray. Before doing the experiment read Chapter 12. Handle the test-tube containing the ^{137}Cs with a glove.

Experiment 6. To determine the half-life of a long lived radionuclide (^{238}U), using a calibrated G-M tube

Theory. ^{238}U has a half-life of $4{\cdot}49\times10^9$ y. The uranium decays by α-emission, but it is in radioactive equilibrium with ^{234}Pa (p. 69), 95% of which decays by emission of β-rays with a maximum energy of 2·33 MeV.

^{90}Y, which is in equilibrium with its parent ^{90}Sr, decays by 100% emission of β-rays of maximum energy 2·27 MeV. The beta energy is therefore virtually equivalent to that of ^{234}Pa. A standardized solution of ^{90}Sr can therefore be used to calibrate a G–M tube for absolute counting of uranium.

Apparatus and materials. Scaler/timer with quench unit, liquid-counting G–M tube, with castle, micro pipette, uranyl (VI) nitrate, (solid), beaker for radioactive waste, standardized solution of ^{90}Sr containing 0·01 μCi per 10 cm^3. (Owing to the high toxicity of ^{90}Sr this solution must be prepared by a qualified teacher from a standardized solution obtained from The Radiochemical Centre.)

Experimental details. Connect up the G–M tube to the scaler and switch on. After a few moments switch on the high voltage and adjust so that it is about 50 V from the start of the plateau of the G–M tube. Switch off the high voltage and withdraw the tube from the castle. Fill the tube with water and replace in the castle. Make sure the tube is dry on the outside, or it may short-circuit. Switch on the high voltage and count the background over 10–20 min.

While this is proceeding weigh out accurately about 1 g of uranyl(VI) nitrate, dissolve in water, and make up to 100 cm^3 in a standard flask. Mix the solution thoroughly.

Empty out the water from G–M tube, rinse with a little propanone (acetone) and dry with a blower.

Pour 10 cm^3 of the ^{90}Sr solution into the G–M tube, return it to the castle and count for 10 min. Pour the ^{90}Sr solution from the G–M tube into a small flask labelled ^{90}Sr. This solution can be retained for future use. Wash the G–M tube thoroughly with strontium carrier (i.e. Sr^{2+} aq.), pouring the washings into the radioactive waste beaker, and then wash with water. Finally, fill with water, replace in the castle, and check that the tube is properly decontaminated by counting the background. If it is still contaminated, rinse with ‘ Decon 90 ’ and check again.

Now empty out the water, dry as before, and pour in 10 cm^3 of the uranium solution. Return to the castle and count for 10 min. Subtract the background from both counts and determine the counting rate for the ^{90}Sr and ^{238}U in counts s^{-1}, expressed to 3 significant figures. Calculate the standard deviation (p. 121) on each of these counts. As the count due to the background will be a small fraction of the count due to source plus background the standard deviation due to the background may be neglected.

The standardized solution of ^{90}Sr will be supplied with a certificate giving the number of μCi cm^{-3} on a particular date. After making any necessary correction for decay and dilution, determine the number of μCi of ^{90}Sr in 10 cm^3 of the solution provided, and hence the number of disintegrations s^{-1}, I_1. (1 μCi is equivalent to $3{\cdot}70 \times 10^4$ dis s^{-1}.)

Let x=counts s^{-1} as determined for the ^{90}Sr.

Let k=efficiency of the counter.

Then
$$k = x/I_1$$

Let y=counts s^{-1} for the ^{238}U.

Let I_2=disintegration rate per 10 cm^3.

Then
$$I_2 = y/k$$

From the measured concentration of the uranium solution determine the disintegration rate per gram of uranium. The formula of uranyl(VI) nitrate is $UO_2(NO_3)_2.6H_2O$.

Calculate the number of atoms of ^{238}U in 1 g of uranium, assuming that the percentage of ^{238}U is 100. This is justifiable as most uranium salts supplied are depleted in ^{235}U. From the equation

$$-\frac{dN}{dt}=\lambda N$$

determine λ and hence the half-life.

Calculate the standard deviation of the measurement as follows: Now

$$t_{\frac{1}{2}} \propto \frac{1}{I_2}=\frac{k}{y}$$

And as

$$k=\frac{x}{I_1}$$

$$t_{\frac{1}{2}} \propto \frac{x}{y} \cdot \frac{1}{I_1}$$

Using the table of standard deviations for arithmetically combined quantities on p. 121 calculate the standard deviation of x/y. A further correction should now be applied similarly for the standard deviation of I_1, as quoted on the certificate of measurement for the ^{90}Sr solution.

Health hazard. There is no health hazard from the uranium and little from the ^{90}Sr because of the small quantity used. However, the filling of the G–M tube must be performed over a tray and the usual precautions taken to monitor the hands. A glove should be used to handle the vessel containing the ^{90}Sr.

Experiment 7. To determine the half-life of a short lived radionuclide by the integral, or 'ratio', method

Theory. The system chosen for this experiment is uranium in equilibrium with its first three decay products (Figure 1.3, p. 8). ^{234}Pa may be extracted from solutions of uranium salts in fairly concentrated hydrochloric acid by use of a suitable organic solvent. Protoactinium and uranium both form solvated anionic complexes $[PaOCl_6]^{3-}$ and $[UO_2Cl_3]^-$ respectively, in solutions with a high concentration of chloride ion. At high $[H^+]$ the undissociated acid H_2PaCl_7 is formed, and this is extracted into the organic phase and solvated; some uranium is also extracted. Thorium does not form undissociated acid complexes under these conditions, and complete separation of ^{234}Pa from the parent ^{234}Th is thus made possible.

Owing to the rapid decay of ^{234}Pa the half-life is most conveniently measured by the integral or 'ratio' method. The theory is fully discussed on p. 79. Two scalers and preferably two timers are necessary. It is helpful to know the approximate value of the half-life so that the second scaler may be started after about two half-lives.

Apparatus. Well-type scintillation counter, or liquid counting G–M tube in castle, with quench unit, 2 scalers and preferably 2 timers, 50 cm^3 separating funnel (stoppered), small beaker, 10 cm^3 measuring cylinder, stop clock.

Materials. Concentrated hydrochloric acid, saturated solution of uranyl(VI) nitrate (100 g per 100 cm^3), 3-methylpentan-2-one (isobutylmethyl ketone), tray with tissues.

Experimental details. If two timers are available arrange the electrical connections so that the pulse from the amplifier is fed to both timers and each timer operates a scaler. If only one timer is available the pulse from the timer must be fed to both scalers and one of the scalers must be switched on at a definite time, say 150 s from the start. Add 6 cm^3 of the uranyl(VI) nitrate solution and 14 cm^3 of concentrated hydrochloric acid to the separating funnel followed by 5 cm^3 of 3-methylpentan-2-one (pentyl ethanoate (amyl acetate) could also be used). Shake thoroughly for a few minutes and allow the phases to separate. Phase separation is very rapid and should be complete in 30 s. Run the lower aqueous phase into the small beaker and the upper organic phase into a specimen tube which fits into the well of the scintillation counter, and has a capacity of 8 cm^3. If a G–M tube is to be used extract with 11 cm^3 of the organic solvent and run this direct into the G–M tube. Immediately switch on the high voltage if the G–M tube is used, though in the case of the scintillation counter the high voltage should already be on. Start the timer without delay. After 150 s switch on the second timer, or if only one timer is to be used start the second scaler. Continue counting for 600 s. Set one of the timers to count for 100 s and determine the background with the uranium solution in the counter. The background will then include any long lived impurities.

The two scalers will give (*a*) the total number of counts recorded over the whole period of the decay, and (*b*) the number of counts from time 150 s to the end of the decay. Subtract the background for 600 s and 450 s respectively from each count. Let x = the counts recorded for the whole period of the decay and x_1 = the counts recorded from time 150 s to the end of the decay. Using Equation (5.20).

$$t_{\frac{1}{2}} = \frac{0{\cdot}301t}{\log_{10}x - \log_{10}x_1}$$

(where $t = 150$ s); calculate the value of $t_{\frac{1}{2}}$ for ^{234}Pa.

The most accurate result is obtained when $x/x_1 = 5$. The half-life of ^{234}Pa is 71 s.

Calculate the statistical error of the result obtained.

Health hazard. There is no appreciable health hazard, but the chemical operations must be carried out over a tray and care must be taken not to spill the concentrated solution of uranyl(VI) nitrate.

Experiment 8. To resolve a decay curve with two components

Theory. When two or more independent radio nuclides are decaying simultaneously this is known as ‘ composite decay ’ (p. 75). Silver has two naturally occurring isotopes, ^{107}Ag and ^{109}Ag. Both have high capture cross-sections (p. 22) for thermal neutrons (^{107}Ag, 24 b and ^{109}Ag, 90 b). The radioactive isotopes produced, ^{108}Ag and ^{110}Ag, decay with half-lives of 2·3 min and 24 s respectively to stable isotopes of Cd. The β-particles emitted have high maximum energies and so can be counted easily with a liquid-counting G–M tube at relatively high efficiency. Owing to the high capture cross-sections they can be prepared with a laboratory neutron source. A convenient neutron source is the americium-beryllium source of strength 0·5–1·0 Ci housed in a large block of paraffin wax, preferably about 0·5 m in diameter (p. 140).

Apparatus and materials. Scaler/timer with quench unit, liquid counting G–M tube, neutron source, stop-clock, 10 cm³ of a saturated solution of silver nitrate.

Experimental details. Pour the solution of silver nitrate into the liquid counting G–M tube, place the tube in the castle and connect to the scaler. Adjust the high voltage so that it is about 50 V above the start of the plateau of the tube, and count the background for 15–20 min. Owing to the high density of a saturated solution of silver nitrate the background will be appreciably lower than with pure water in the tube.

Switch off the high voltage, remove the tube from the castle and pour the silver nitrate solution into a test-tube with a piece of thread tied round the rim. Cork the test-tube, lower it into one of the holes in the neutron assembly and leave for 15 min. The neutron source should be within easy reach of the scaler, but NOT in the same room.

Set the scaler to count for 15 s. Mark the glass of the stop-clock at 20, 40 and 60 s respectively. It will be necessary to count for 15 s at 5 s intervals. This requires some practice and is best done by two students, one to record the results and the other to switch on and off and re-set the scaler. Set the second hand of the stop-clock to 55 s and have pencil and paper ready.

Withdraw the silver nitrate solution from the neutron source and bring it as quickly as possible to the G–M tube. Pour the solution into the G–M tube, replace the tube in the castle and switch on the high voltage. Start the stop-clock with the second hand at 55 s, and when the second hand reaches 0 switch on the scaler. Note the reading at 15 s, re-set and switch on again at 20 s. Continue recording 15 s counts for about 10 min. Subtract the background per 15 s from each reading, and plot a graph of the counting rate against time taking the first count as zero time. Remember that the counts are at 20 s intervals. Towards the end there will be wide fluctuations. Draw a smooth curve through the points, to even out the fluctuations. Take readings *from this curve* at 30 s intervals and plot the log of these readings against the time, or use semi-log paper.

The first part of this graph will be a curve which will become a straight line after about 2·5 min (see Figure 5.3, p. 76). Produce the straight line backwards to zero time. From this straight line determine the half-life of ^{108}Ag.

Subtract the readings of the straight line graph from the upper curve, starting at time 0, at 20 s intervals and plot these differences on the same graph. This will give another straight line, but much steeper. From the second straight line determine the half-life of ^{110}Ag.

When the experiment has been completed pour the silver nitrate solution back into the stock bottle, and wash the G–M tube thoroughly with pure water.

Health hazard. There is no health hazard from the radioactive silver owing to the short half-life. However, concentrated silver nitrate will produce severe blackening of the hands and care must be taken not to spill any on the skin. Wash the hands thoroughly after handling vessels containing the silver nitrate solution. The hazard from the neutron source depends on the nature of the source (see p. 162 for precautions in the use of neutron sources). The radioactivation with the neutron source must be carried out by a qualified instructor or technician.

Further work. Other suitable elements for studying composite decay are rhodium and bromine. The bromine must be separated from an organic bromide by Szilard–Chalmers reaction (see Chapter 11). Rhodium metal may be irradiated with neutrons and counted with an end window G–M tube. Rhodium has one natural isotope, ^{103}Rh. This reacts with a thermal neutron to give two nuclear isomers (p. 96) of ^{104}Rh. The excited state decays by internal conversion ($t_{\frac{1}{2}}=4{\cdot}4$ min), to the ground state ($t_{\frac{1}{2}}=43$ s). The capture cross-section for the shorter lived isomer is 131 b as against 11 b for the longer lived isomer. Rhodium is to be preferred to silver as it is easier to count (20 s counts at 30 s intervals would be suitable) and in consequence gives a better graph.

If no neutron source is available it is possible to make up an artificial mixture to study composite decay. A suitable mixture would be ^{234}Pa separated from uranium in concentrated HCl by extraction with 3-methyl-pentan-2-one, and ^{208}Tl separated from natural thorium in concentrated HCl (with a little bromine water to oxidize the thallium to covalent Tl(III)), by extraction with 2-methylethoxypropane (di-isopropyl ether). Concentrated (50%) solutions of both uranium and thorium as nitrates should be used and 5 cm^3 of each extract should be mixed and poured into the G–M tube. 20 s counts at 30 s intervals would be suitable. $^{234}Pa, t_{\frac{1}{2}}=72$ s; ^{208}Tl, $t_{\frac{1}{2}}=3{\cdot}1$ min.

Experiment 9. To plot a γ spectrum

Theory. See Chapter 9, p. 132 and Figures 9.5 and 9.6.

Apparatus. Scaler/timer with single channel pulse height analyser, scintillation counter with high resolution crystal.

Materials. ^{137}Cs about 1·0 μCi, ^{60}Co about 1·0 μCi.

Experimental details. The caesium source can be conveniently prepared by withdrawing a suitable quantity from the stock bottle with a micro syringe, and discharging into a stoppered specimen tube. (This should be done by a qualified teacher or technician—the source once prepared may be used repeatedly.) Place the source on top of the scintillation counter, switch on the high voltage, and set the discriminator to 0·66 MeV with the channel integral. Switch on the ratemeter, or start the scaler and slowly increase the high voltage until a rapid increase in counting rate occurs. This is produced by the ^{137}Cs photopeak.

Now set the channel width to 2% and increase or decrease the high voltage slightly until a maximum counting rate is achieved. The analyser is now set correctly to read the energy of the pulses in MeV.

Turn the discriminator to the lowest value, i.e. about 0·1 MeV, set the timer to 30 s and do a series of counts at 0·02 MeV intervals, continuing until well past the photopeak of 0·66 MeV. Plot the results on a graph with counts per 30 s on the y axis and energy in MeV on the x axis. Measure the width of the photopeak at half height in MeV, and express this as a percentage of 0·66, which is a measure of the resolution of the scintillation counter (p. 134).

Repeat the experiment with ^{60}Co and measure the height of the 1·33 MeV peak and also the height of the dip between the two peaks, taking the base line as zero count rate. This is another measure of resolution (p. 135).

Health hazard. There is no appreciable health hazard, but the tubes containing the ^{137}Cs and ^{60}Co must be held by the top and they should be carried to the scintillation counter in a small beaker for safety.

Extension. Other interesting γ-spectra for study are ^{110m}Ag and ^{131}I.

Experiment 10. To plot a β-absorption curve and to determine the maximum β-energy

Theory. See Chapter 9, p. 128, and Figure 9.1.

Apparatus. Scaler/timer with quench unit, end-window G–M tube in suitable tube holder (see Appendix 7, p. 255), set of aluminium absorbers of known thickness. Beta source of 0·1–1·0 μCi activity.

Experimental details. ^{90}Sr β-sources of the required activity are generally available. Alternatively a ^{32}P source can be prepared by evaporating a suitable aliquot of $Na_3{}^{32}PO_4$ stock solution on an aluminium planchet under an infrared lamp.

Connect the G–M tube to the quench unit, switch on the scaler/timer, and adjust the high voltage so that it is about 50 V higher than the start of the plateau of the G–M tube. Determine the background over at least 5 min.

Place the β-source on a shelf in the tube holder at such a distance from the tube that it gives not more than 10 000 counts min^{-1} and record a 1 min count. Insert an aluminium absorber of thickness about 10 mg cm^{-2} on the shelf immediately below the tube and repeat the count. Do a series of counts using increasing thicknesses of aluminium until the counting rate becomes reasonably constant, slightly above background. With a ^{32}P source the thickness should be increased in steps of about 20 mg cm^{-2}, but with ^{90}Sr double this increase would be suitable. As the counting rate falls the time of count should be increased, but it will be impractical to maintain constant statistical accuracy.

Plot a graph of log counting rate on the y axis against thickness of absorber on the x axis and determine the range of the β-particles from the point where the graph flattens out. Determine the energy by the use of Figure 9.2, p. 129.

If time is short the half-thickness may be determined. In this case it is only necessary to make about 8–10 counts. With ^{32}P these should be in steps of about 20 mg cm^{-2} continuing until the count rate is reduced to about one-third of the initial count and with ^{90}Sr steps of 30 mg cm^{-2} would be sufficient. Plot a log graph similar to the previous one. The graph should be a straight line. Determine the half-thickness from the graph and the maximum beta energy by the use of Figure 9.3.

Health hazard. There is no appreciable health hazard.

Experiment 11. To measure the resolving time of a counter assembly

There are three principal methods in common use:

(*a*) the method of paired sources;

(*b*) successive additions of a nuclide of long half-life;

(*c*) extrapolation of a known decay curve.

Theory. Resolving time is explained in Chapter 8, p. 122.

Apparatus. Scaler/timer with G–M tube, or scintillation counter (with a G–M tube it is preferable to use a quenching unit), aluminium, copper or stainless steel disc of about 1–2 cm diameter depending on the diameter of the tube window, infrared lamp, micro pipette (10 or 50 mm^3 capacity), tray with tissues, bottle for liquid waste.

Materials, ^{137}Cs (10 μCi in 0·2 cm^3) for (*a*) and (*b*), ^{128}I, as prepared in Experiment 15 is suitable for (*c*).

Experimental details. (1) This method is most suitable for an end window G–M tube. Fix the tube in the tube holder or castle, set the quenching unit to 800 μs, or any other value to suit the tube, and place a metal tray on a shelf about 1 cm from the tube window. The metal tray must have a circle drawn on it, or better a metal ring soldered to it, of diameter equal to the metal disc supplied. This circle or ring should be immediately beneath the tube window. Count the background for 20 min.

Cut the metal disc in two and put a smear of silicone oil on each half. Using the micro pipette add a small volume of the ^{137}Cs solution on to the disc and evaporate to dryness under the lamp. Repeat the additions until about 1 μCi has been added. The total volume required would be about 20 mm^3. This should be added in two or three lots. It is of the utmost importance to prevent the drop spreading, hence the use of silicone oil. Repeat the operation with the second half of the disc, adding approximately the same quantity of the ^{137}Cs solution.

Handling the half disc with forceps place it on the metal tray so that it occupies half of the circle and count for 300 s. Now place the second half of the disc under the counter so that it occupies the other half of the circle. Take great care not to move the first half disc. Repeat the count for the same time. Finally, remove the first half disc and repeat the count.

Let A be the true count s^{-1}, A' the observed count s^{-1}, τ be the resolving time and B the background,
then

$$A = \frac{A'}{1 - A'\tau} \tag{E.11.1}$$

since $A'\tau$ is the time in seconds during which the counter is inoperative. As $A'\tau$ is small this expression can be simplified to:

$$A = A'(1 + A'\tau)$$

Now for the first count:

$$A_1 = A'_1(1 + A'_1\tau)$$

and similarly:

$$A_2 = A'_2(1 + A'_2\tau)$$

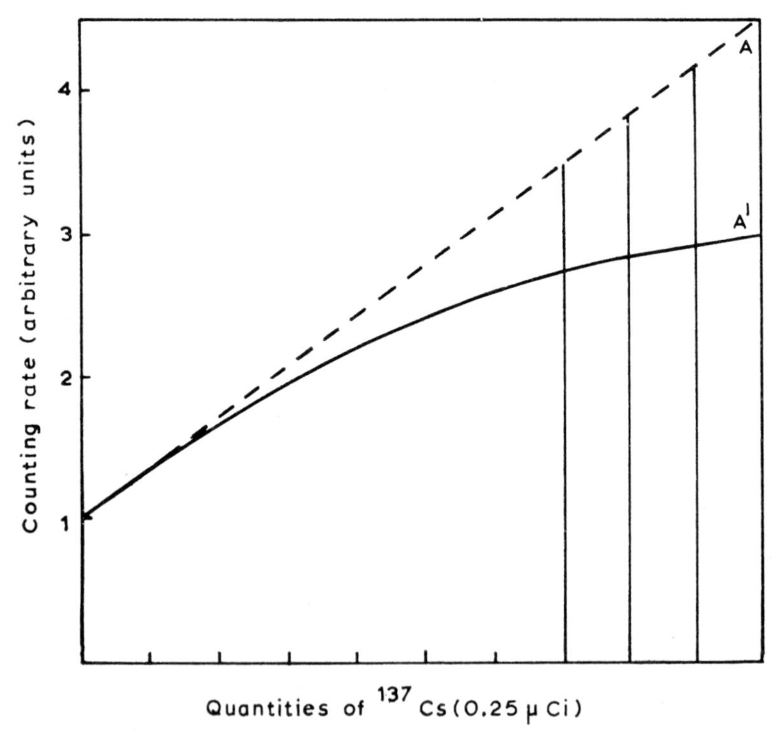

Figure E.11. (*a*) The graph shows how to measure resolving time by successive additions of a long lived radionuclide. The lower curve shows the observed counting rate. The resolving time may be determined from the corresponding readings of the lower and upper curves and the use of Equation E 11.1.

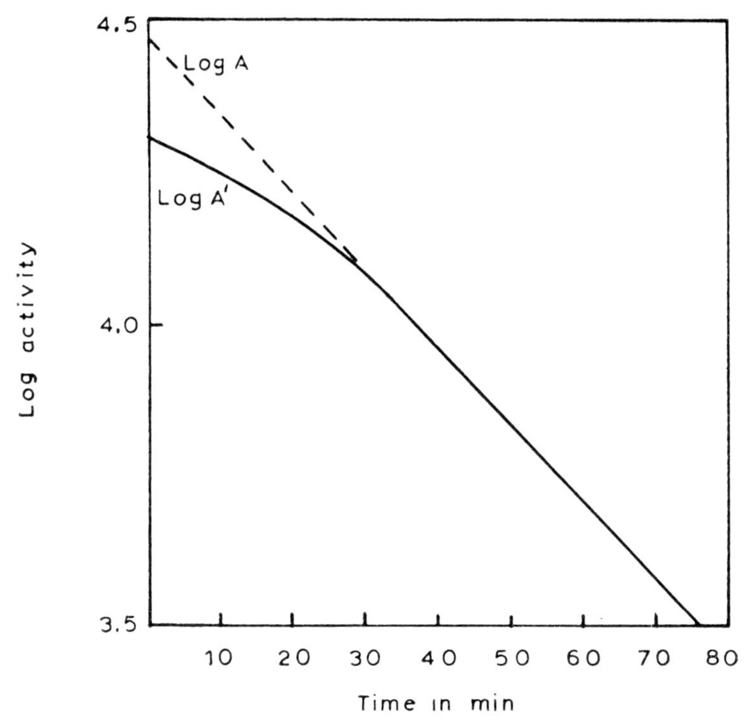

Figure E.11 (*b*) The graph shows the method of measuring resolving time by use of a radionuclide with a reasonably short half-life such as ^{128}I. At high counting rates the graph of log activity (A′) is a curve, due to lost counts. By measuring the values of log A′ and log A from the extrapolated portion of the curve the values of A′ and A may be used in Equation E.11.1 to determine the resolving time.

and:

$$A_{12} = A'_{12}(1 + A'_{12}\tau)$$

where A'_{12} is the observed count with both sources together.

Also:

$$A_{12} + B = A_1 + A_2$$

(where B = the background over 300 s)
since A_{12}, A_1 and A_2 each include the background.

Hence:

$$A'_{12}(1 + A'_{12}\tau) + B = A'_1(1 + A'_1\tau) + A'_2(1 + A'_2\tau)$$

extracting τ gives:

$$\tau = \frac{A'_1 + A'_2 - A'_{12} - B}{{A'_{12}}^2 - {A'_1}^2 - {A'_2}^2}$$

(2) This is the best method for a scintillation counter, using a 15% gate. Under these conditions the resolving time will be about 20 μs.

Add in succession 0·25 μCi quantities of the ^{137}Cs solution to a small specimen tube for a well type counter, or to a small beaker for a solid crystal. After each addition take a count for 300 s. Take about 10 readings. Plot a graph of the counting rate against each addition (see Figure E.11 (*a*)). Take several readings of A and the corresponding reading of A' from the graph, and calculate τ by the use of Equation E.11.1. The most accurate result will be obtained when the difference is a maximum.

(3) This method could be used with any type of counter, but is ideal for a liquid counting G–M tube using a nuclide such as ^{128}I of fairly short half-life (25·0 min), emitting a high energy β-particle. The ^{128}I is extracted into 10 cm^3 of aqueous solution after irradiation of 50–100 cm^3 of iodoethane with a neutron source of about 1 Ci (see Experiment 15). After determining the background a series of counts should be made at intervals of five minutes over about three half-lives, i.e. 75 min. The initial count rate should be about 20 000 min^{-1}. The initial counting time could be about 1 min, lengthening to 2 min and 4 min as the counting rate falls, to maintain reasonable statistical accuracy.

Plot a graph of log counting rate against time. The initial part of the graph will be a curve due to lost counts, but the final part should be a straight line (see Figure E.11 (*b*)). Extrapolate the straight part of the graph to $t=0$ and hence determine log A. Now calculate τ by the use of Equation E.11.1 and the values of A and A' at $t=0$.

Health hazard. The ^{137}Cs solution should be in a small bottle fitted with a serum cap and stand in a small lead pot. The student must handle the bottle with forceps over a tray. The discs used in (1) must be handled with forceps. After use the half discs should be washed with a little CsCl solution as carrier and the solution discarded to liquid waste. Alternatively, the two sources could be prepared by a qualified instructor or technician and lacquered, when they may be used repeatedly.

The ^{137}Cs solution used in (2) must be added to the specimen tube or beaker over a tray, and the tube handled with tongs or forceps. After use the tube or beaker must be washed with carrier and the carrier solution discarded to liquid waste.

There is no health hazard in (3) owing to the short half-life of ^{128}I, but the irradiation of the iodoethane with the neutron source must be carried out by a qualified instructor or technician.

Experiment 12. An introduction to liquid scintillation counting

Theory. See Chapter 7, pp. 111 to 117.

Apparatus. Liquid scintillation counting head, scaler/timer with pulse height analyser, supply of phials, micro-pipette of capacity 50 μl (mm^3).

Materials. Liquid scintillator, nitrogen cylinder, 5 cm^3 pipette with syringe attached, silicone oil about 30 c/s, ^{14}C labelled hexadecane and

^{3}H labelled hexadecane (both obtainable from The Radiochemical Centre—a few μCi only required), tetrachloromethane.

Experimental details. Unscrew the cap of the liquid scintillator bottle and bubble nitrogen slowly through the liquid for a good ten minutes. The presence of oxygen reduces the efficiency of counting. While this is in progress check the silicone oil level in the scintillator head unit and add more if required. Connect up the cooling coil of the photomultiplier tube to the cold tap and turn on the water (unless a refrigerated unit is in use). Connect the scintillator head unit to the scaler/timer and switch on the HV.

Next pipette 5 cm^3 of liquid scintillator into each of four of the special phials provided (this operation should be done in a subdued light). Mark the phials 1 to 4 unless they are numbered and insert No. 1 into the light-tight lock. Rotate the knob and bring the phial into contact with the photomultiplier tube†. The phial should be left in contact for at least 5 min to ‘ dark adapt ’ before counting. If the instrument has a variable gain amplifier this should be set to 250 or 500. Set the lower discriminator so that it is about one tenth of the maximum, and the upper at ‘ integral ’, and adjust the HV until counts are observed. Take a series of 30 s counts at intervals of 50 V on the HV scale until the background begins to increase rapidly. It is not advisable to exceed 1500 V.

(1) *To determine the efficiency of counting* ^{14}C *and* ^{3}H

Take the tube of ^{14}C labelled hexadecane and place this in a little hot water if necessary. The melting point of hexadecane is 18·5°C. Transfer 30 μl of the hexadecane by means of the micro pipette (or such a quantity as will give about 25 nCi) into phial No. 2. Withdraw phial No. 1 from the scintillation head, after allowing it to drain for about one minute to avoid unnecessary loss of silicone oil. Replace with phial No. 2 ‘ dark adapting ’ as before. Repeat the counting procedure exactly as with phial No. 1. Correct each count of the ^{14}C specimen (phial No. 2) for background and calculate the efficiency of the counter for ^{14}C. An example should make the method clear.

Lower discriminator 0·1 MeV. Upper discriminator integral. Counts/30s. Phial 2 contained 24·6 nCi as calculated from the volume taken, the density of hexadecane, and the specific activity. This is equivalent to 27306 dis/30 s.

HV	Background Phial No. 1	^{14}C Phial No. 2	^{14}C corrected for background	% efficiency
900	24	23050	23026	84·3
950	39	24061	24022	88·0
1000	274	25628	25354	92·9
1050	2071	28250	26179	95·9

This experiment shows the advantage of liquid scintillation counting as a method of determining ^{14}C. For instruments with a variable gain amplifier it is preferable to fix the HV and determine the counting rate

† The technique for inserting the phials varies with the make of the scintillator head unit. The maker's instructions should be followed.

against amplifier gain. Both amplifier gain and HV may be varied to obtain optimum counting conditions.

Repeat the experiment for tritium using the ^{3}H labelled hexadecane, and putting a given volume in phial No. 3. The efficiency for counting tritium will be much lower owing to the low maximum energy of the β-particles from tritium (0·018 MeV compared to 0·155 MeV for ^{14}C).

(2) *To plot the beta spectrum of ^{14}C and ^{3}H individually*

Place phial No. 2 in the scintillator head and set the lower discriminator to 0·1 MeV or at 5 V according to the instrument. Set the gating circuit to 10% and the HV to a setting with a low background such as 900 V in the above example. Carry out a series of 10 s counts at intervals of 0·02 MeV or 1 V depending on the instrument. Plot the counts obtained on the y axis against MeV or volts on the x axis and draw a smooth curve through the points. Repeat the experiment with phial No. 3, ^{3}H. Graphs such as Figure 7.7 should be obtained. (P. 112).

(3) *To count ^{14}C and ^{3}H together in a mixture by the channels ratio method*

For this experiment the instructor should prepare a mixture of ^{14}C and ^{3}H in phial No. 4 and give this to the student as an 'unknown'.

This problem is complicated by the fact that the total integrated count will be the sum of the counts for ^{3}H and ^{14}C and no instrument yet devised can differentiate between the two. However, the problem can be solved in a very simple way. First set the HV to a suitable value such as 1000 V and insert phial No. 3 containing the known amount of tritium. Set the discriminator to 0·1 MeV or 1 V and the gating circuit to integral and determine the number of counts over a suitable time such as 100 s. Calculate the efficiency. Now raise the discriminator level until no counts are recorded, i.e. it is beyond the end of the tritium spectrum. Record the value of the discriminator. Replace phial No. 3 by phial No. 2 and count the ^{14}C with the discriminator first at 0·1 MeV or 1 V and then in the position where no counts are recorded with tritium. Determine the efficiency for ^{14}C with the discriminator at 0·1 MeV and also the ratio of the two counts for ^{14}C.

Now replace phial No. 2 with phial No. 4 (the unknown mixture), and determine the total count with the discriminator first in the lower position (0·1 MeV) and then in the upper position. The second count is that given by ^{14}C alone. From these measurements calculate the amount of ^{14}C and ^{3}H in the mixture. An example will make the method clear.
HV=1000 V. Counts per 100 s.

Tube content	Amount in nCi	Counts with discriminator setting 0·7 MeV		0·1 MeV	
		Gross	Net	Gross	Net
Pure scintillator	0	27	—	608	—
^{3}H	52·7	51	24	45928	45320
^{14}C	24·6	46105	46078	85850	85242
Mixture of ^{14}C and ^{3}H	?	30978	30951	83217	82609

The count of 24 for ^{3}H with discriminator setting 0·7 MeV can be disregarded. The count for the mixture at 0·7 MeV can be regarded as due to the ^{14}C alone. Hence the proportion of ^{14}C counted at 0·1 MeV.

$$=82\,609\times\frac{30\,951}{46\,078}=55\,489$$

The count due to the ^{3}H

$$=82\,609-55\,489=27\,120$$

Hence the amount of ^{14}C

$$=\frac{55\,489}{85\,242}\times 24{\cdot}6=\mathbf{16{\cdot}0\ nCi}$$

and the amount of ^{3}H

$$=\frac{27\,120}{45\,320}\times 52{\cdot}7=\mathbf{31{\cdot}5\ nCi}$$

(4) *To study the effect of quenching*

The principle and methods have been fully discussed in Chapter 7, pp. 113 and 116. A suitable quenching agent to use is CCl_4, because very small quantities are required to produce considerable quenching. Add a known amount of hexadecane labelled with ^{14}C to 5 cm^3 of scintillator and determine the count. Calculate the efficiency; hexadecane produces no quenching. Add successive quantities, e.g. 10 μl, of CCl_4 and determine the counting rate and hence the efficiency. Plot the efficiency against the volume of CCl_4 added. Similar curves can be constructed for any substance and a quench correction applied provided the amount of the substance present is known.

CCl_4 is also convenient for the channels ratio method. The experiment is conducted exactly as described above except that two counts are made on each specimen, one in each channel as described on p. 116, where the advantages of this method have been fully discussed.

Experiment 13. To determine the thermal neutron flux in a paraffin block containing a neutron source

Theory. The most accurate and convenient method for measuring a low neutron flux is to irradiate a known weight of an element of high capture cross-section, and accurately known half-life, for a known time, and to count it with a calibrated counter. Gold, as a solution of $HAuCl_4$, is the most suitable element. Its capture cross-section is 98 b, it has one natural isotope ^{197}Au and its half-life is accurately known (2·696 d).

The counter, preferably a well-type scintillation counter, or a liquid counting G–M tube, must be calibrated by use of a gold reference solution.

The neutron flux will consist of epithermal and thermal neutrons. Epithermal neutrons have energies somewhat in excess of thermal energies. The capture cross-section for gold is high for low energy epithermal and for thermal neutrons, but very low for fast neutrons. Thermal neutrons are absorbed by cadmium foil. To obtain the correct thermal flux two measurements must be made, one with the gold solution surrounded with a tube of cadmium and one without the cadmium tube.

The thermal flux is determined by subtracting the count with the cadmium tube from the count without the cadmium.

The neutron flux is determined from the equation

$$f = I_0 / N\sigma(1 - e^{-\lambda t}) \qquad \text{(E.13.1)}$$

where f is the flux, I_0 the disintegration rate per second at the end of the irradiation, σ the capture cross-section, λ the decay constant for ^{198}Au, and t the time of irradiation in seconds. (The student is referred to Chapter 5, p. 73 for the derivation of the above equation.)

Apparatus and materials. Scalar/timer with well-type scintillation counter (or liquid counting G–M tube), neutron source, stop watch, tray with tissues, 100 cm^3 standard flask, 1 and 2 cm^3 syringe pipettes, micro syringe, solution of gold 'chloride' ($HAuCl_4$) of known concentration, a standardized reference solution of gold (about 5 μCi).

Experimental details. A well-type scintillation counter is much to be preferred to a G–M tube because the solution of gold chloride can be kept permanently in a closed tube and used repeatedly.

Before carrying out the experiment the counter must first be calibrated. A reference solution of gold as $KAu(CN)_4$ may be obtained from the Radiochemical Centre. These standardized solutions are prepared at 4-monthly intervals. 5 μCi would be sufficient. A certificate will be supplied giving the activity in μCi at a particular time. A known fraction of the solution containing about 1 μCi of activity should be diluted to 100 cm^3 in a standard flask and an aliquot, e.g. 2 cm^3, of this solution counted with the scintillation counter, setting the pulse height analyser to 0·42 MeV† with a 10% channel width. 50 000 counts should be taken and the time recorded at the mid-point of the count. The background must also be recorded, preferably over 1000 s.

The counting rate (A) must be corrected for dead time if necessary (see Chapter 8, p. 122 and Experiment 11) and for decay (A_0), i.e. at the time stated on the certificate. The efficiency of the counter is the ratio of $(A_0/I_0) \times$ dilution factor where I_0 is the decay rate of the standard.

The efficiency of scintillation counters should be checked periodically using a long lived source such as ^{137}Cs, counting with the setting of the pulse height analyser as for gold.

† 0·42 MeV is the energy of the principal γ-ray from ^{198}Au.

To determine the neutron flux weigh accurately about 1 g of $HAuCl_4$ into a specimen tube of the same size as that used to calibrate the counter. Dissolve the $HAuCl_4$ in a total volume of water of about 2 cm^3 and cork the tube securely.

Place the tube in the position where it is required to measure the neutron flux and leave it for at least 10 days (with a 0·5 Ci neutron source). The time of irradiation must be recorded to the nearest hour.

When continuing the experiment set the pulse height analyser to 0·42 MeV with a 10% channel width and measure the background over 1000 s, if time permits. Withdraw the tube from the neutron source and count it with the scintillation counter with the setting as above. A correction for decay is unnecessary if the count is complete within half an hour of withdrawing the tube. With a neutron flux of 10^3 n cm^{-2} s^{-1} and a weight of gold chloride as recommended, a count of about 20 000 should be recorded in 1000 s.

From the known efficiency of the counter and the counting rate observed, determine I_0 and the flux by use of Equation E.13.1. This will give the combined thermal and epithermal flux.

To determine the thermal flux alone repeat the experiment, but surround the tube of gold chloride with a tube of cadmium and obtain the thermal flux by difference.

Calculate the standard deviation of your measurement.

Health hazard. The bottle containing the 5 μCi of ^{198}Au must not be handled with the fingers and should be kept in a lead pot. The diluting of the active solution must be carried out over a tray. The operation with the neutron source must be performed by a qualified instructor or technician. A glove should be worn when mixing the active solution in the standard flask.

Note for use with a G–M tube. In this case the gold solutions must be diluted to a suitable volume for the tube.

Experiment 14. Neutron activation analysis

(*a*) *To detect impurities in a compound by reactor irradiation*

Theory. As an example of this technique a crude specimen of 'technical grade' cerium oxide (CeO_2) has been chosen. This product is almost certain to contain lanthanum and samarium and probably traces of praseodymium, neodymium, europium and gadolinium. Many of the natural isotopes of the lanthanides have high capture cross-sections for thermal neutrons and in consequence traces are relatively easy to detect by neutron activation analysis. A small amount of the oxide is irradiated in a reactor for a suitable time and a complete γ-spectrum plotted covering the range from 0·09 to 1·6 MeV. The best analysis will be obtained with a multichannel analyser and solid state detector, but several of the impurities will be detected by use of a good quality sodium iodide crystal and single channel spectrometer. The impurities present can be determined from the energy peaks and the corresponding half-lives.

TABLE E.14.1. *Nuclear characteristics of cerium and likely impurities*

Isotope	Abundance (%)	Activation cross-section (b)	Half-life of isotope produced	Principal γ-energies (MeV)
^{140}Ce	88·5	0·27	32·5 d	0·145
^{142}Ce	11·1	0·10	33 h	0·29, 0·72, 1·10
^{139}La	99·9	8·2	40·2 h	0·33, 0·49, 0·82, 0·92, 1·60
^{141}Pr	100	11·0	19·2 h	1·57
^{146}Nd	17·2	0·31	11·1 d	0·091, 0·53
^{152}Sm	26·7	37	46·2 h	0·10
^{151}Eu	47·8	670†	9·3 h	0·122, 0·84
		3360	13 y	0·34, 1·41, 0·96
^{153}Eu	52·2	220	16 y	0·12, 1·01, 1·28
^{158}Gd	24·9	1·0	18 h	0·36

Apparatus and materials. Scaler/timer with single channel pulse height analyser and high resolution sodium iodide crystal, or multichannel analyser and solid state detector, cerium oxide (CeO_2, technical grade).

Experimental details. Irradiate 0·1 g of cerium oxide in a neutron flux of 10^{12} n cm^{-2} s^{-1} for 1 hour and then plot the complete gamma spectrum. Count the principal peaks for a suitable time and repeat at daily intervals. It should not be necessary to make a correction for background as this will be very small when using the necessary narrow ' gate '. Plot decay graphs and determine the half-lives. From the energies of the peaks and the corresponding half-lives determine the impurities present.

Health hazard. There is no appreciable health hazard, but the tube containing the cerium oxide must be handled with tongs, and kept in a lead pot when not in use.

(*b*) *To determine the percentage of manganese in a specimen of ' Dural ' by radioactivation analysis, using a neutron source*

Theory. Most specimens of ' Dural ' contain 90–92% of aluminium, 4–5% of copper, 0·7% of silicon, 0·8% of magnesium and small quantities of other metals. The nuclear characteristics of the isotopes of the principal elements present, which could be activated appreciably with a neutron source, are:

Isotope	Abundance (%)	Activation cross-section (b)	Half-life of the isotope produced	Principal γ-energies (MeV)
^{26}Mg	11·2	0·003	9·5 min	0·834 (30%), 1·015 (70%)
^{27}Al	100	0·21	2·27 min	1·78 (100%)
^{30}Si	3·0	0·11	2·62 h	no γ
^{55}Mn	100	13·3	2·58 h	0·845 (99%), 1·81 (24%), 2·12 (15%)
^{63}Cu	68·1	3·0	12·8 h	0·51‡ (15%), 1·34 (0·6%)
^{65}Cu	31·9	0·56	5·1 min	1·04 (9%)

† Producing ^{152m}Eu. ‡ Positron annihilation energy.

From a study of this table it should be clear that only Al and Cu can interfere with the determination of Mn and this interference can be eliminated by allowing the specimen to decay for 20 min and by use of a discriminator. (The student should check this for himself.)

The experiment can be carried out with a neutron source giving a flux of 10^3 n cm^{-2} s^{-1} and irritating the specimen of Dural for about 12 h. The counter must be calibrated for manganese by irradiating a known weight of a pure specimen of manganese, or a manganese salt, in the same flux for the same time and counting this specimen.

Apparatus and materials. Scaler/timer, with discriminator and sodium iodide crystal, neutron source, 2 polythene tubes, 'Analar' manganese(II) sulphate, or specimen of pure manganese, sheet of 'Dural'.

Experimental details. Irradiate about 10 g of manganese(II) sulphate ($MnSO_4.4H_2O$) (or about 5 g of manganese), and 10 g of 'Dural', both weighed accurately, in polythene tubes for 12 h, or overnight. It is important that the neutron flux should be the same, or accurately known, for both samples. (For the measurement of neutron flux see Experiment 13). Set the discriminator at 0·8 MeV and count the background for 1000 s. Withdraw the manganese sulphate and the 'Dural' from the neutron source and record the time. Take this as zero time.

Tip the manganese sulphate into a small beaker of approximately the same cross-sectional area as the piece of 'Dural', stand it on top of the counter and count for 1000 s. Record the time at 500 s and correct the counting rate to zero time, after subtracting the background. Determine the counting rate per gram of manganese and so calibrate the counter.

Now count the 'Dural' also for 1000 s, recording the time at 500 s, and correct the counting rate to zero time as for the manganese sulphate.

From these two observations and the known weight of the specimen of 'Dural', determine the percentage of manganese in the 'Dural'. It will be necessary to make a correction if the neutron flux was not the same for the two irradiations.

Calculate the standard deviation of the measurement.

Health hazard. There is no appreciable health hazard, but the operation with the neutron source must be performed by a qualified instructor or technician.

Experiment 15. To determine the percentage retention of ^{128}I in the organic phase after neutron irradiation of iodoethane

Theory. This is the original Szilard–Chalmers reaction and the theory is fully discussed in Chapter 11, p. 149.

Apparatus. Scaler/timer with quench unit, liquid counting G–M tube in lead castle, neutron source with an output of at least 2×10^5 n cm^{-2} s^{-1}, 50 cm^3 separating funnel, tray with tissues.

Materials. Iodoethane (25 cm^3), 10% solution of sodium sulphite (about 50 cm^3) containing a little potassium iodide, phase separating filter paper.

Experimental details. The percentage retention decreases with the concentration of iodine scavenger present. The iodoethane should preferably be redistilled before use, or shaken with sulphite, washed and dried and stored in an amber bottle to avoid photodissociation. Weigh out 0·01 g of iodine and dissolve this in 25 cm^3 of the iodoethane. Calculate the molar fraction of iodine. Irradiate the iodoethane in the neutron assembly for about 2 h before use.

Connect up the G–M tube and adjust the HV to a suitable value. Count the background for at least 5 min. Withdraw the iodoethane from the neutron assembly and without delay pour about 10 cm^3 into the G–M tube. (Remember that this operation must be done over a tray and NOT with the tube in the castle; also remember to switch off the HV before removing the G–M tube from the castle.) Return the tube to the castle, switch on the HV and count for 5 min. Start a stop clock half way through the count.

While the count is proceeding extract the free iodine from the remaining iodoethane by shaking twice with the sulphite solution. Pour the contents of the G–M tube back into the stock bottle and refill the tube with the iodoethane from the separating funnel and count again for 5 min. Stop the stop clock half way through the count. Wash the contents of the tube and any residual iodoethane from the separating funnel twice with water and filter into the stock bottle through phase separating filter paper.

Correct each count for dead-time (see Chapter 8, p. 122) and background and then correct the second count for decay, i.e. the time recorded on the stop clock, by use of the equation

$$\ln A_0 = \ln A + \lambda t \quad (\lambda \text{ for } {}^{128}\text{I} = 2{\cdot}77 \times 10^{-2}\ \text{min}^{-1})$$

Express the second count as a percentage of the first count. Advanced students should determine the statistical accuracy of their result.

Health hazard. There is no health hazard owing to the small amount of activity involved and the short half-life of ^{128}I, but all operations with the neutron source must be carried out by a qualified instructor or technician.

Further work. It is instructive to repeat this experiment at a series of different molar fractions (m.f.) of scavenger iodine, starting at 10^{-5} m.f. and continuing until a saturated solution of iodine in iodoethane is obtained. Draw a graph of percentage retention in the organic phase on the y axis against m.f. of iodine on the x axis and comment on this graph. Different organic iodides both aliphatic and aromatic could be used.

It is also instructive to investigate the effect of adding aniline (aminobenzene) to the iodoethane and determining the percentage retention. This should be reduced by the presence of aniline since the following reaction occurs:

$$C_2H_5{}^{128}I + C_6H_5NH_2 \rightarrow (C_6H_5NH_2C_2H_5)^+ + {}^{128}I^-$$

This is the Menschutkin reaction and the activation energy is high. The newly formed $C_2H_5{}^{128}I$ has the iodine in an excited state and this favours the reaction. The aniline can be removed from the iodoethane by shaking with dilute hydrochloric acid.

Experiment 16. To determine the distribution coefficient for iodine between water and 1,1,1-trichloroethane

Theory. Iodine dissolved in 1,1,1-trichloroethane is labelled with ^{128}I, obtained by Szilard Chalmers reaction using a neutron source, and equilibrated with water at a known temperature. The ratio of the concentration of the iodine in the two layers is determined by counting with a suitable G–M tube. If a neutron source is not available ^{131}I can be used.

Apparatus. Neutron source with an output of about 10^6 neutrons s^{-1}, scaler/timer with quench unit, liquid counting G–M tube, or scintillation counter, stoppered separating funnel to hold 200 cm^3, 2 corked 100 cm^3 conical flasks. 10 and 5 cm^3 syringe pipettes, 25 cm^3 measuring cylinder.

Materials. Iodoethane, containing iodine at about 10^{-5} molar fraction, 100 cm^3, 1,1,1-trichloroethane containing iodine, about M/20, 20 cm^3, 100 vol. hydrogen peroxide, bench ammonium molybdate, bench dilute sulphuric acid (M/1), petroleum spirit (60/80), sodium sulphite solution, about 3%, tray lined with tissues.

Experimental details. Irridiate the iodoethane containing iodine in two 50 cm^3 polythene bottles placed as near as possible to the neutron source for about 3 h, or overnight. Pour the active iodoethane into the 200 cm^3 separating funnel and add 10 cm^3 of the sodium sulphite solution. Shake for about 30 s. Run off the iodoethane (lower layer) into a conical flask and retain for further use. Purify as described in Experiment 15. Add about 20 cm^3 of the solution of iodine in 1,1,1-trichloroethane to the separating funnel, followed by 10 cm^3 of the dilute sulphuric acid, two drops of the ammonium molybdate solution, and 10 drops of 100 vol. hydrogen peroxide. Shake for one minute. The sulphite is rapidly oxidized to sulphate and the iodide to iodine (ammonium molybdate acts as a catalyst), which then exchanges with the iodine in the 1,1,1-trichloroethane layer, leaving about 1% of the activity in the aqueous layer. Run out the 1,1,1-trichloroethane (lower layer) into a 100 cm^3 conical flask, and add about 15 cm^3 of water. Stopper the conical flask, and shake vigorously in a thermostat at 25°C for about 15 min. At the end of this time withdraw a suitable aliquot of the aqueous layer, and count with the liquid β-counter. Take particular care to avoid including any drops of the 1,1,1-trichloroethane solution. Record the time at the mid point of the count. Six minutes counting should be sufficient, and this should give 2000–3000 counts with a G–M tube, which is preferable for ^{128}I.

While the aqueous layer is being counted withdraw 5 cm^3 of the 1,1,1-trichloroethane layer and mix with 15 cm^3 of petroleum spirit. This is to produce a liquid with a density approximately equal to water, and so avoid a correction for differences in counter efficiency when counting liquids of different density. Pour out the aqueous layer from the counter and replace with a similar aliquot of the 1,1,1-trichloroethane–petroleum spirit mixture. Count the layer as before, and record the time at the mid point of the count.

Notes on the calculation. Correct the organic layer for decay by using Equation 5.17, Chapter 5, p. 74.

Multiply the result by 4 (to allow for dilution of the organic layer) and divide by the counting rate of the aqueous layer. The result should be accurate to 2–3%. The published figure is 90 : 1 at 25°C. Advanced students should calculate the statistical accuracy of their result.

Health hazard. There is no health hazard, and no problem about decontamination owing to the short half-life of ^{128}I. The irradiation with the neutron source must be performed by a qualified instructor or technician.

Note.—If a neutron source is not available then ^{131}I may be used. 0·1 μC is sufficient if a scintillation counter is available, but at least 1 μC must be used with a G–M tube. In this case the ^{131}I, which is supplied carrier free in sodium thiosulphate, should be added to the 10 cm^3 of dilute sulphuric acid in the separating funnel after the 20 cm^3 of iodine in 1,1,1-trichloroethane has been added. The remaining sequence should be followed exactly, except that there is no need to correct for decay.

After the experiment the aqueous layer can be discarded to waste. The 1,1,1-trichloroethane, including that mixed with the petroleum spirit, should be shaken with sulphite solution and the active iodine extracted as iodide should be discarded to liquid waste. Alternatively, the 1,1,1-trichloroethane layer can be retained and used for a second student. It should be possible to use this solution four or five times by adding a little 1,1,1-trichloroethane to make up for that lost in counting. All apparatus used should be decontaminated by rinsing with potassium iodide solution, and tested with a suitable monitor before being washed in the usual way.

Health hazard. ^{131}I is an isotope of medium toxicity. Students should not handle more than 1·6 μCi. It is advisable to use rubber or polythene gloves when shaking the separating funnel. Do NOT handle G–M tubes with gloves.

Experiment 17. To investigate the solubility of lead chloride in hydrochloric acid solutions using ^{212}Pb as a tracer, and to determine the value of the stability constant for the $PbCl_4^{2-}$ complex ion

Theory. ^{212}Pb is present in all thorium compounds. By reference to Figure 1.2, p. 7, it will be seen that, with the exception of ^{228}Ra and ^{228}Th all the decay products have $t_{\frac{1}{2}} < 4d$. A freshly prepared thorium compound will have been separated from the decay products, but with the exception of ^{228}Ra these products will reach radioactive equilibrium with ^{228}Th in less than one month, and the total activity will decay initially with the half-life of ^{228}Th. This activity will pass through a minimum of 60% of the original activity after 5 y, and then will slowly increase, due to the growth of ^{228}Ra, reaching full activity again effectively after about 40 y (a brief reference to this has already been made in Experiment 2).

The most convenient method of extracting ^{212}Pb is by use of a thorium ' cow ' (Figure E.17), so-called because it can be ' milked ' at intervals of 2–3 d almost indefinitely. The thorium hydroxide (preferably an aged specimen) is continuously giving off ^{220}Rn. The release of the radon is helped by a moist atmosphere. ^{220}Rn, ^{216}Po, and ^{212}Pb are all positively charged, due to loss of electrons on expulsion of the α-particle. They are therefore attracted to the negatively charged platinum wire. ^{212}Pb collects on the wire in radioactive equilibrium with the thorium. If the apparatus is kept in a corner of the laboratory it acts as a permanent source of ^{212}Pb and its decay products.

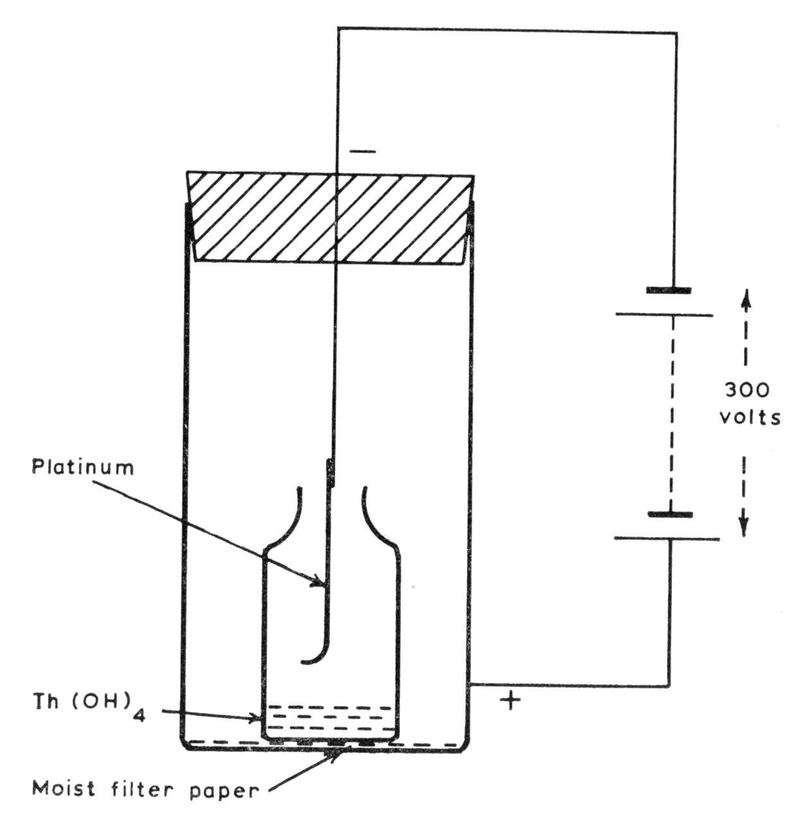

Figure E.17. The thorium ' cow '. The $Th(OH)_4$ is continuously giving off ^{220}Rn, the atoms of which are + charged due to loss of electrons. They collect on the − charged platinum wire and rapidly decay to ^{216}Po and so to ^{212}Pb. After $2\frac{1}{2}$ days equilibrium will be established. The ^{212}Pb and its decay products may be dissolved off the wire and the ^{212}Pb and ^{212}Bi used as tracers. If a copper wire with a point is used instead of the platinum wire it makes a useful α-source for cloud chambers.

The activity may be dissolved off the wire into a solution of a lead salt, thus providing a tracer for lead. Bismuth, thallium and polonium will also be present in minute traces, but only the ^{212}Bi need be considered here. The lead must be precipitated as chloride. To prevent ^{212}Bi being carried down with the precipitate, bismuth is added as a carrier.

The ^{212}Bi will grow again in the lead chloride (transient equilibrium, Appendix 4, p. 248), and reach maximum activity after 3·8 h. ^{212}Bi gives off β-rays of 2·25 MeV energy, whereas the β-rays from ^{212}Pb are only 0·4 MeV energy. If β-counting is intended the samples should be counted 3–4 h after the precipitation of the lead chloride. Gamma counting can be carried out immediately, because the ^{212}Pb emits a high proportion of γ-rays of 0·24 MeV energy. Even so it is advisable to use a standard of known lead concentration to allow for the growth of ^{212}Bi with its daughter ^{208}Tl, and for the decay of the ^{212}Pb.

The precipitated lead chloride, labelled with ^{212}Pb, is dissolved in solutions of hydrochloric acid of known concentration and the concentration of lead ions determined by counting aliquots, comparing the counts with those obtained using the standard solution.

Apparatus. Scaler/timer with well-type scintillation counter, or liquid counting G–M tube in castle with quench unit, special apparatus for collecting ^{212}Pb (Figure E.17), rack of corked test-tubes (9), 2 cm^3 pipette with syringe, centrifuge, tray with tissues.

Materials. Solution of lead nitrate of known concentration (about 10%), solutions of hydrochloric acid, 0·1, 0·2, 1, 2, 4, 6, and 10M (need not be exact standards), dilute nitric acid (2M).

Experimental details. Remove the wire from the apparatus (with the cork) and test the activity by bringing it close to the window of the G–M tube of a monitor. 500 counts s^{-1} should be recorded. Heat 2 cm^3 of 2M-HNO_3 in a small beaker, after adding 2 drops of the solution of lead nitrate and 2 drops of a 1% solution of bismuth nitrate. When the solution is boiling gently insert the platinum wire. 2–3 min heating in the nitric acid should remove most of the activity from the wire (test it at intervals with the monitor). The solution should be evaporated to near dryness to remove excess HNO_3, and the residue dissolved up again in 20 cm^3 of the lead nitrate solution. 2 cm^3 of this active solution must be retained in a corked tube as a reference standard.

Precipitate the lead in the remaining 18 cm^3 as chloride by addition of dilute hydrochloric acid (about 0·5 M). Centrifuge and wash two or three times with the minimum amount of water. Add about 4 cm^3 of water to one of the test-tubes and 4 cm^3 of each of the solutions of hydrochloric acid to each of the remainder. Transfer some of the washed lead chloride to each tube. There must be sufficient lead chloride to leave a permanent precipitate after shaking. Start by adding a very small quantity, about 0·05 g, to the tubes containing water and the dilute acids, but considerably more lead chloride will be required for the more concentrated acid. The tubes should be corked and after thorough shaking placed in a thermostat, or bowl of water at 20°C for about 30 min. The precipitate settles very rapidly. 2 cm^3 should be removed from each tube with the syringe pipette and counted with the scintillation counter at known times, with the discriminator set to 0·24 MeV and gate width 15%. At least 2000 counts should be taken if possible, but this may be difficult with the more dilute solutions. The 2 cm^3 of the active lead nitrate solution as a standard should be counted at the beginning of the series of counts and again at known time intervals, and a graph of log activity plotted against time.

If it is necessary to use a liquid β-counter the 2 cm^3 of saturated solution must be diluted to a suitable volume for the tube and all the solutions left to stand for 3–4 h after the original precipitation of lead chloride.

Calculate the concentration (molarity) of Pb^{2+} in each tube by comparison with the standard graph, and plot a graph of $[Pb^{2+}]$ against molarity of the hydrochloric acid.

Notes on the calculation. The initial fall in the graph is caused by the 'common ion effect'. The solubility of the lead chloride increases with increase in $[Cl^-]$ due to complexing. A range of complex ions are formed up to $PbCl_6^{4-}$ in the most concentrated acid solutions. Over the range 3–6 M HCl the principal complex ion is $PbCl_4^{2-}$. The stability constant of this complex may be determined as follows:

The solubility of lead chloride in pure water should first be calculated and hence the solubility product k_s. Let x stand for the total $[Pb^{2+}]$ and y for the total $[Cl^-]$. $k_s = xy^2 = 4x^3$.

The stability constant

$$K = \frac{[Pb^{2+}][Cl^-]^4}{[PbCl_4^{2-}]} = \frac{k_s y^2}{[x - (k_s/y^2)]} \text{ mol}^4 \text{ dm}^{-12} \qquad \text{(E.17.1)}$$

neglecting the Cl^- in the complex in comparison to the free Cl^- since Cl^- is present in large excess.

Equation (E.17.1) can be further simplified when $y > 3$ (k_s is very small compared to y^2) to:

$$K \approx \frac{k_s y^2}{x}$$

Calculate the values of K for values of $y = 3$, 4, 5, and 6 M.

Exact agreement cannot be expected because $[Cl^-]$ is not equal to the total acid molarity and other complex ions are present, but quite reasonable values should be obtained in the region of 3–6 M.

$$k_s \text{ at } 20°C = 1{\cdot}86 \times 10^{-4} \text{ mol}^3 \text{ dm}^{-9}$$

(Note that $k_s = 4[Pb^{2+}]^3$ NOT $[Pb^{2+}]^3$, since for each Pb^{2+} there are 2 Cl^-.)

Health hazard. The breathing should be restrained when the cork is removed from the thorium 'cow'. There is very little danger owing to the short half-life of the products. It is advisable to carry out the work using the ^{212}Pb over a tray.

Experiment 18. To plot a solubility curve for lead iodide using ^{212}Pb as a tracer

Theory. ^{212}Pb is obtained from a thorium 'cow' as in Experiment 17. Bismuth nitrate is added in excess to lower the specific activity of the bismuth compared to that of the lead. Both are precipitated as iodides.

Bismuth iodide is appreciably less soluble than lead iodide. A saturated solution is prepared at the required temperature and a known volume withdrawn and counted with a scintillation counter, with the discriminator set to 0·24 MeV, using a 10 or 20% gate. This practically eliminates the count from the bismuth. The concentration of lead iodide is determined by comparing the counting rate of the saturated solution with that of a standard.

Apparatus. Scaler/timer with well-type scintillation counter (or liquid counting G–M tube), special apparatus for collecting ^{212}Pb (Figure E.17), 2 cm^3 pipette fitted with syringe, centrifuge, boiling tube, 110°C thermometer, large beaker (1 dm^3 capacity), 25 cm^3 standard flask.

Materials. Solid lead nitrate, solid bismuth nitrate, potassium iodide solution (10%), dilute sulphuric acid and dilute nitric acid (2 M), tray with tissues.

Experimental details. Dissolve about 0·1 g of lead nitrate (weighed accurately), and about 0·4 g of bismuth nitrate, in a few cm^3 of dilute nitric acid and warm on a hot plate. Withdraw the active platinum wire from the thorium ' cow ' and stir gently in the hot solution. Withdraw the wire at intervals and test with a portable monitor. The activity should fall to about 10% of its initial value in a few minutes.

Make up the solution of labelled lead and bismuth nitrates to 25 cm^3 in a standard flask, mix well, and withdraw 2 cm^3 by pipette and put in a corked tube for retention as a standard. Precipitate the remainder of the solution, in the 25 cm^3 standard flask, as iodide by adding about 8 cm^3 of 10% potassium iodide solution. At the same time precipitate the 2 cm^3 of standard solution of labelled lead nitrate as sulphate by adding dilute sulphuric acid. Centrifuge the precipitate and wash it. This effectively removes the bismuth since bismuth sulphate is soluble. Centrifuge also the precipitated iodides, wash them well with the minimum of water, and slurry into a boiling tube with about 20 cm^3 of water. Immerse the boiling tube in a beaker of water which can be heated or cooled to any desired temperature.

To make sure that the solution is saturated heat the boiling tube some 10 degrees above the desired temperature and then allow to cool slowly in the beaker until the temperature of the measurement is reached. Withdraw 2 cm^3 of the saturated solution and place it in a tube for counting. Care must be taken to avoid drawing any of the precipitate into the pipette. If desired a short length of glass tubing filled with glass wool may be attached to the pipette. This precaution should not really be necessary as the precipitate settles very rapidly. At temperatures above room temperature the pipette must be warmed before inserting into the saturated solution. Wash any precipitate forming in the pipette into the tube. Some precipitation of lead iodide will occur in the tube on cooling, but this will have no appreciable effect when a well-type scintillation counter is used. Record the time of each count, and take at least 10 000 counts.

Count the lead sulphate standard at suitable intervals of time in between counts of the saturated solution. These counts could be taken while the saturated solutions are being extracted. Plot a graph of log activity of standard against time. This will show transient equilibrium, but the counting rate of the bismuth can be much reduced by suitable use of a discriminator of gating circuit.

The activity of the saturated solution must be compared with the activity of the standard at the appropriate time. Since the weight of lead ions in the standard is known, the solubility of lead iodide can be determined from the relative activities. Draw a graph of solubility of lead iodide in g per 100 g of solution against temperature. It is not easy to work at temperatures much above 60°C.

Note on the use of a G–M tube. If a scintillation counter is not available it is possible to do the experiment with a liquid counting G–M tube. Assuming the capacity of this tube to be about 10 cm^3, 8 cm^3 of water should be added to each of the 2 cm^3 specimens of saturated solution. Any precipitated lead iodide will be redissolved. These solutions should all be counted 3–4 h after the original precipitation of the iodides to allow maximum activity of the ^{212}Bi to form in the lead iodide. If it is desired to use a standard the precipitation of lead as sulphate must be omitted. Instead, precipitate the lead as chloride using M HCl, cool the solution well in ice water and centrifuge. Re-dissolve the precipitate in 10 cm^3 of concentrated hydrochloric acid. This procedure should remove nearly all the ^{212}Bi originally present, but the results will not be as accurate as with the scintillation counter.

Health hazard. There is no appreciable health hazard, though the breathing should be restrained when opening the thorium 'cow'. The work with the ^{212}Pb should be conducted over a tray.

Experiment 19. To determine the stability constant for the silver ammine complex ion

Theory. Silver salts which are insoluble in water dissolve in ammonia solution due to the formation of complex ions. This experiment makes use of ^{110m}Ag as a tracer to determine the formula for the complex ion and the value of the stability constant at ammonia concentrations between 1·0 and 0·05 M. Silver nitrate solution of known concentration is labelled with ^{110m}Ag and is converted to silver chloride which is washed and dried. Small quantities of this silver chloride are added to centrifuge tubes (boiling tubes could be used) containing solutions of ammonia at varying concentrations. The tubes are placed in a thermostat and stirred. Aliquots of each solution are withdrawn, after centrifuging if necessary, and counted with a scintillation counter. A similar aliquot of the labelled silver nitrate solution is also counted as a standard and the concentration of silver in each tube containing ammonia determined by comparison of the counting rate with that of the standard solution.

Apparatus. Well-type scintillation counter with scaler/timer, 5 or 6 centrifuge tubes of 15 cm^3 capacity, or boiling tubes, fitted with paraffined corks or glass stoppers, thermostat at 25°C with rack to hold tubes, electric stirrer to fit tubes, 2 cm^3 pipette with syringe, sintered glass filter-crucible, porosity 3 or 4, centrifuge (preferably).

Materials. Silver nitrate, silver nitrate solution containing ^{110m}Ag, M/1 ammonia solution, tray with tissues.

Experimental details. Prepare a standard solution of silver nitrate containing about 1·5 g in 100 cm^3 and add 3 μCi of the active silver nitrate solution before making up to the mark. If necessary, make a correction for the weight of silver nitrate carrier added to the standard solution from the active solution. Retain 10 cm^3 of the standard solution, transfer the remainder to a beaker and precipitate the silver as chloride by the addition of 10 cm^3 of 2 M hydrochloric acid. Place the beaker in the dark and leave for some hours to coagulate. Filter the silver chloride on the sintered glass filter using a water pump and avoid strong light. Wash the precipitate well with water, finally with a little acetone, and dry in the oven at 110°C for about 20 min.

Standardize the solution of ammonia and prepare solutions of 0·5, 0·25, 0·125, 0·063 M by dilution. Add about 5 cm^3 of each molarity including the M/1 solution to separate centrifuge tubes, or boiling tubes with about 0·2 g of the active silver chloride, and label the tubes. The quantity of silver chloride can be estimated; there is no need to weigh it. Place the tubes in the thermostat, stir for about 5 min, cork and leave for a day. Centrifuge the tubes. This is not absolutely essential as the precipitate generally settles out on the bottom, but if a centrifuge is available it is a wise precaution. If a centrifuge is not available, attach a short length of glass tubing filled with glass wool to the end of a pipette with a piece of rubber tubing, as a filter. Use a fresh filter tube for each solution and place the filters in the radioactive waste bin. Withdraw 2 cm^3 by pipette (or a larger volume to suit the capacity of the scintillation counter) from one of the tubes, and count for at least 10 000 counts, having set the discriminator to give the maximum ratio of counts of specimen to background. Return the solution to the tube from which it was withdrawn and stir for about 5 min. While this is in progress count the next solution, and repeat the procedure for all five solutions. Repeat the counting procedure after stirring and centrifuging to make sure that equilibrium has been reached. If there is a significant difference in the counting rates, further stirring will be necessary until the counting rate becomes constant. Finally count the standard solution. Correct the counting rates for lost counts where necessary and for background, and determine the concentration of silver in each tube in terms of the counting rate compared to that of the standard solution. No correction for decay need be applied provided that the final counts of unknowns and standard are performed within three days.

Health hazard. All operations involving active solutions must be carried out over a tray. Rubber or polythene gloves should be worn. The active silver chloride must be placed on a tray in the oven, and a warning sign displayed. The residues should be placed in the residues bottle, precipitated as silver chloride and either stored or disposed of as solid waste. The vessels used should be rinsed out with ammonia and checked on the scintillation counter. If necessary they must be rinsed with a special radioactive decontaminating detergent, such as ' Decon '.

Notes on the calculation.

$$Ag^+ + xNH_3 \rightleftarrows Ag(NH_3)_x^+$$

$$K = \frac{[Ag^+][NH_3]^x}{[Ag(NH_3)_x^+]}$$

$$\log [Ag(NH_3)_x^+] = \log [Ag^+] + x \log [NH_3] - \log K \qquad (E.19.1)$$

Now since the silver was added in the form of silver chloride, $[Cl^-]=[Ag]_t$ where $[Ag]_t$=total concentration of silver ions, both free and combined in the complex.

The solution must be saturated with silver chloride, so:

$$[Ag^+]=\frac{K_s}{[Cl^-]}=\frac{K_s}{[Ag]_t} \qquad (E.19.2)$$

where K_s is the solubility product of silver chloride at the temperature of the thermostat.

Assuming that $[Ag]_t$ in the most dilute ammonia solutions is about 10^{-3} mol dm^{-3}$=[Cl^-]$, and, taking $K_s=10^{-10}$ mol^2 dm^{-6}, $[Ag^+]$ will be about 10^{-7} mol dm^{-3}. A negligible error will be made by assuming that

$$[Ag(NH_3)_x^+]=[Ag]_t \qquad (E.19.3)$$

and neglecting the concentration of the free Ag^+.

Substituting in Equation (E.19.1), from (E.19.2) and (E.19.3),

$$\log [Ag]_t=\log \frac{K_s}{[Ag]_t}+x \log [NH_3]-\log K$$

or

$$2 \log [Ag]_t=x \log [NH_3]+\log (K_s/K) \qquad (E.19.4)$$

$[NH_3]$ cannot be calculated exactly until the value of x is known, but as the ammonia is in considerable excess the fraction in the complex can be neglected to a first approximation. The graph of $\log [Ag]_t$ against $\log [NH_3]$ will not be quite straight because of this approximation, but the slope $(\frac{1}{2}x)$ will give the value of x. The correct concentrations of free ammonia can now be calculated, and a fresh graph plotted of $\log [Ag]_t$ against $\log [NH_3]$. The intercept at $\log [NH_3]=0$ gives the value of $\log (K_s/K)$ and the value of K, the stability constant of the complex ion, may be determined from the known value of K_s at the temperature of the thermostat. ($K_s=1{\cdot}10\times10^{-10}$ mol^2 dm^{-6} at 20°C and $1{\cdot}56\times10^{-10}$ mol^2 dm^{-6} at 25°C.)

Complete agreement with theory cannot be expected for the following reasons:

(1) The ionic strength of the solution is not zero and this will increase the solubility of silver chloride.

(2) The concentration of free ammonia is less than the calculated value because some of the ammonia is present as $NH_3.H_2O$ and as NH_4^+.

(3) x varies with the concentration of ammonia and the temperature. The value obtained from the graph is unlikely to be a whole number.

Further work. Repeat the experiment, but precipitate silver bromide instead of silver chloride. Using the value of the stability constant previously determined calculate the solubility product of silver bromide.

Experiment 20. To investigate the equilibrium between iodide ions and molecular iodine in aqueous solution

Theory. Iodine is sparingly soluble in water, but dissolves readily in solutions of potassium iodide. You are required to determine the formula of the complex ion formed and the value of the stability constant at a known temperature, using ^{131}I as a tracer.

Apparatus. Scaler/timer with well-type scintillation counter, 4 stoppered boiling tubes, 100 cm³ separating funnel, 2 × 2 cm³ pipettes with syringes, 2 × 5 cm³ graduated pipettes with syringes, 100 cm³ corked flask, 25 cm³ measuring cylinder, tray with tissues, rubber or polythene gloves.

Materials. Standard potassium iodide solution, about 0·1 M, containing 1·5 μCi of ^{131}I per 50 cm³ (50 cm³ only required) labelled A, 1,1,1-trichloroethane, dilute sulphuric acid (w), 10-volume hydrogen peroxide, ammonium molybdate solution.

Experimental details. Take 20 cm³ of solution A (by measuring cylinder) and transfer to the tap funnel. Add by measuring cylinder 5 cm³ of dilute (M) sulphuric acid, 20 cm³ of 1,1,1-trichloroethane, 5 cm³ of 10 volume hydrogen peroxide, and 2 or 3 drops of ammonium molybdate solution. Ammonium molybdate is a powerful catalyst for the oxidation of iodide ions by hydrogen peroxide (Experiment 16). Shake the separating funnel thoroughly to extract the iodine into the 1,1,1-trichloroethane, and then run out the violet-coloured layer of 1,1,1-trichloroethane into the 100 cm³ flask. Separate the 1,1,1-trichloroethane layer as completely as possible. Label the flask ' solution B—active ' and cork securely.

Next add to each of the four stoppered test tubes 5 cm³ of the standard active potassium iodide solution A followed by 5, 3, 2·5 and 2 cm³ of the active solution B. The graduated pipette is convenient for this. Make up the 1,1,1-trichloroethane layer to 5 cm³ by adding 0, 2, 2·5, 3 cm³, respectively, of inactive 1,1,1-trichloroethane. Stopper these tubes securely, using springs to retain the stoppers, if available, and shake in the thermostat. Leave for a good twenty minutes with frequent shaking. A mechanical shaker is useful.

Withdraw 2 cm³ lots from each layer by pipette and count with the scintillation counter. When withdrawing the bottom layer (1,1,1-trichloroethane) blow gently with a syringe to expel any of the aqueous layer, before filling the pipette. Take particular care not to spill active solution on to the scintillation counter.

Count 2 cm³ of the original solution A. The count of solution A will enable you to relate counts to concentration.

Health hazard. All work with active solutions must be performed over a tray and gloves must be worn except when handling pipettes, or bottles of inactive solutions. Radioactive waste must be placed in the flasks provided and all apparatus decontaminated after the experiment using potassium iodide and sulphite followed by ' Decon '. It would be preferable to work under a fume hood.

Notes on the calculation. The partition coefficient for iodine between 1,1,1-trichloroethane and water is 90 at 25°C.

Let a = count rate in counts/min/2 cm³ for the aqueous layer,
b = count rate in counts/min/2 cm³ for the 1,1,1-trichloroethane layer,
c = count rate in counts/min/2 cm³ for the potassium iodide solution A,
d = concentration of solution A in g dm^{-3},
K_d = partition coefficient.

The concentration of the various species present may be expressed initially as counts min^{-1}.

(1) Mass of I^-, I_2 and I_3^- in 1 dm³ of the aqueous layer $\equiv a$.

(2) Mass of $I_2 \equiv b/K_d$.

(3) Mass of I^- and $I_3^- \equiv a-(b/K_d)$.

(4) Mass of I_2 combined in $I_3^- \equiv a-(b/K_d)-c$.

(5) Mass of $I^- \equiv a-(b/K_d)-\frac{3}{2}[a-(b/K_d)-c]$
(Since the total weight of iodine in I_3^- must be $\frac{3}{2}\times$ the weight of I_2 in the complex)
$=\frac{3}{2}c-\frac{1}{2}[a-(b/K_d)]$.

Multiply 2, 4 and 5 by (d/c) to express the concentrations of I_2, I^- and I_3^- in g dm^{-3}. Convert these quantities into mol dm^{-3}, remembering that I_3^- is determined in terms of I_2 combined as I_3^-, and hence divide by 2×127 NOT 3×127.

Tabulate the results and calculate the four values of the stability constant expressed as:

$$K=\frac{[I_3^-]}{[I^-]\,.\,[I_2]}\ \text{mol}^{-1}\ \text{dm}^3$$

Experiment 21. To determine the solubility product of silver ethanoate at a known temperature and to investigate the effect of ionic strength on solubility

Theory. Silver ethanoate (acetate) is a sparingly soluble salt. In this experiment the solubility is determined in pure water, in solutions of silver nitrate and sodium acetate, and also in solutions of sodium nitrate and copper(II) nitrate. It is convenient to do this by use of ^{110m}Ag as a tracer, and to determine the concentration of silver ions in the various solutions by comparing the counting rates of these solutions with that of a standard solution of silver nitrate containing the same proportion of ^{110m}Ag.

Apparatus. Scaler/timer with scintillation counter (preferably well-type), eight 15 cm³ centrifuge tubes, or test tubes fitted with paraffined corks or glass stoppers, thermostat at 25°C with rack to hold tubes, electric stirrer or shaker, centrifuge, 5, 2 and 1 cm³ graduated pipettes fitted with syringes, tray with tissues, 3 × 1 cm³ pipettes would be advisable, standard flask 50 cm³.

Materials. Silver nitrate, silver nitrate solution containing ^{110m}Ag, sodium ethanoate, solutions of sodium ethanoate, sodium nitrate and copper(II) nitrate, 1·5 M (5 cm³ of each solution will be required), bottle labelled ' active residues '.

Experimental details. Prepare a standard solution of silver nitrate about 1·5 M. Pipette 10 cm³ of this solution into a beaker, and add 2 μCi of ^{110m}Ag. If necessary correct for the weight of silver nitrate carrier added to the standard solution from the active solution. Mark this solution 'A'. Withdraw by pipette 2 cm³ of solution A, and dilute to 50 cm³. Mark this solution ' B ', and retain for counting as the standard. Dissolve about 5 g of sodium ethanoate crystals in about 5 cm³ of water, transfer to a 15 cm³ centrifuge tube, and add 5 cm³ of solution A. Stir and centrifuge to collect the precipitated silver ethanoate. Wash this thoroughly with several small quantities of water, stirring and centrifuging each time. If a centrifuge is not available collect and wash the silver ethanoate on a small buchner funnel. Place the filter paper in the radioactive waste bin. Pour the washings into the active residues bottle. Wash the purified silver ethanoate into a flask of about 100 cm³ capacity using a total volume of about 40 cm³ of water. Place the flask in the thermostat and stir with an electric stirrer, or shake with a shaker, but in this case the flask must be securely corked. Alternatively, the flask can be left in the thermostat for a day or two with an occasional shake to form the saturated solution. Aliquots should be withdrawn at intervals, filtered and counted with the scintillation counter. When the counting rate becomes steady the solution is saturated.

Transfer 2·5 cm³ of the saturated solution of silver ethanoate to each of the eight tubes. Use the 5cm³ graduated pipette for this, having first rinsed it thoroughly with inactive silver nitrate and washed with water. Label the tubes 1–8. Add to tubes 1, 2 and 3, 0·1, 0·2 and 0·4 cm³ of the 1·5 M labelled silver nitrate solution 'A' respectively and the same volumes of the 1·5 M sodium ethanoate to tubes 4, 5 and 6. Add 0·4 cm³ of the 1·5 M sodium nitrate to tube 7 and 0·4 cm³ of the 1·5 M copper nitrate to tube 8. Use the 1 cm³ graduated pipette for this operation and rinse thoroughly before use. Withdraw some solid silver ethanoate from the flask containing the saturated solution using a glass spatula (a glass rod flattened at the end and bent slightly is suitable), and add to tubes 7 and 8. The solubility of silver ethanoate will be increased in the solutions containing the sodium nitrate and copper nitrate, whereas it will be depressed and silver ethanoate precipitated in the solutions containing a common ion. Tubes 1–6 should reach equilibrium quickly, whereas it will be necessary to stir or shake tubes 7 and 8. All the tubes should be stoppered and left to reach equilibrium in the thermostat for at least a day.

When continuing the experiment centrifuge each tube and return to the thermostat. If a centrifuge is not available use a filter tube (see Experiment 19, p. 201). Withdraw 2 cm³ of the solution from each tube in turn taking great care not to draw up any of the solid, and determine the counting rate with the scintillation counter, setting the lower gate or discriminator at a suitable value to reduce the background, and the upper gate integral. Repeat the counting procedure for the saturated solution, and also for the standard solution B. Take at least 10 000 counts and correct for background and dead time where necessary. No correction is necessary for decay provided all the counts are done within three days. Determine the concentration of silver ions in each tube by comparison of the counting rate with that of the standard solution B. Rinse the 2 cm^3 pipette very carefully between each operation, and start with the solution of lowest activity, i.e. that containing 0·4 cm^3 of sodium ethanoate (1·5 M) solution.

Health hazard. All operations involving active solutions must be carried out over a tray. Rubber or polythene gloves should be worn. All active residues and washings must be placed in the residues bottle, precipitated as chloride and either stored, or disposed of as solid waste in quantities not exceeding 1 μCi. The vessels should be rinsed with ammonia and checked on the scintillation counter. If necessary they should be rinsed with a special radioactive decontaminating detergent, such as ' Decon '.

Notes on the calculation. The $[Ag^+]$ in each solution is determined as explained above.

The $[Ac^-]$ (ethanoate (acetate) ion concentration) can be determined in the following way. In the saturated solution of silver ethanoate and in tubes 7 and 8 $[Ac^-]=[Ag^+]$. In tubes 1, 2 and 3, $[Ac^-]=[Ag^+]-[NO_3^-]$. In tubes 4, 5 and 6, $[Ac^-]=[Ag^+]+[Na^+]$.

Ionic strength (μ). This is defined as $\mu=\frac{1}{2}\Sigma c_i.z_i^2$, where c_i=concentration of each ion and z_i is the charge carried by the ion. Example: to find the ionic strength of a solution 0·03 M with respect to lead chloride and 0·1 M with respect to copper(II) nitrate.

$$[Pb^{2+}]=0{\cdot}03\ \text{M},\quad [Cl^-]=0{\cdot}06\ \text{M},\quad [Cu^{2+}]=0{\cdot}1\ \text{M},\quad [NO_3^-]=0{\cdot}2\ \text{M}$$

$$\mu=\tfrac{1}{2}\Sigma 0{\cdot}03\times4+0{\cdot}06\times1+0{\cdot}1\times4+0{\cdot}2\times1$$

$$=\tfrac{1}{2}\times0{\cdot}78=0{\cdot}39$$

The solubility, and hence the solubility product, increases with increase in ionic strength due to the overall interionic attractions. The following equation connects solubility product and ionic strength:

$$\log k_s=k\sqrt{\mu}+A$$

where k and A are constants.

Tabulate the results showing the concentration of each ionic species in the various solutions with the values of k_s and μ. Plot a graph of $\log k_s$ against $\sqrt{\mu}$ and extrapolate to $\mu=0$. This will give the thermodynamic or ideal value of K_s.

Experiment 22. To determine the transport number of the Ag^+ in silver nitrate solution

Theory. Transport number is defined as the fraction of the current carried by one of the ions in a binary electrolyte. This experiment uses the ionic migration method with a ' Hittorf ' H tube and silver nitrate solution labelled with ^{110m}Ag. The transport number of the Ag^+ is measured by determining the mass of silver which has moved out of the anode compartment, divided by the mass of Ag^+ discharged. These masses are determined in terms of the activity of the tracer. The mass discharged is normally measured with a silver voltameter. In this experiment the need for a voltameter is eliminated by counting aliquots of both anode and cathode liquids, and also an aliquot of the solution before electrolysis. The decrease in the anode count is proportional to the mass of Ag^+ which has moved out of the anode compartment, since the silver dissolving from the anode is inactive. The decrease in the cathode count is proportional to the mass of Ag^+ discharged at the cathode—the mass which has moved in from the anode compartment. The sum of these two differences is therefore proportional to the mass of Ag^+ which has discharged, i.e. it is proportional to the total current carried.

$$\text{The transport number of the } Ag^+ = \frac{\text{the decrease in the anode count}}{\text{the decrease in the anode + cathode counts}}$$

Apparatus. Scaler/timer with well-type scintillation counter, or liquid counting G–M tube in castle with quench unit, Hittorf H tube (Figure E.22) with silver electrodes, 120 V battery, 2 cm³ pipette (10 cm³ for G–M tube) fitted with syringe, 2 × 50 cm³ conical flasks, vessel labelled ' active waste ', tray with tissues, small tray or large beaker to transport the flasks with active solution to a balance.

Materials. Silver nitrate solution, about M/100, containing 1 μCi of ^{110m}Ag (5 μCi for G–M tube) per 100 cm³ (the total volume required will depend on the dimensions of the apparatus).

Experimental details. First examine the electrodes to make sure that the copper to silver soldered joints are intact and also that the silver rods are properly sealed into the glass tubes: Araldite is a suitable watertight adhesive. Clamp the apparatus securely over the tray leaving sufficient clearance to place the small conical flasks under the taps. Open the central tap and pour the silver nitrate solution into the apparatus, retaining 3–4 cm³ of the solution for counting as a standard. Make sure that all air bubbles are removed from the tap. Connect up the battery and switch on, noting the time. It is important to switch on as soon as possible after inserting the electrodes into the solution so as to avoid appreciable exchange occurring between the silver electrodes and the solution. Leave the apparatus undisturbed for 1–2 h. It is most important that the apparatus should not be moved during the run. With the concentration of silver nitrate recommended and the PD suggested the current will not exceed 2 mA and the heating effect will be negligible.

While the electrolysis is proceeding count 2 cm³† of the silver nitrate solution, setting the discriminator as for Experiment 21 and recording at least 10 000 counts (A), and weigh the two small conicals, marked + and −, to the nearest 0·01 g.

After 1–2 h close the central tap, switch off and disconnect the battery. Without delay run about 30 cm³ of the solution from each limb into the respective flask and weigh it (w_+, w_-). For safety the flask should be carried to the balance on a small tray, and the base wiped with a tissue. Now withdraw 2 cm³† from each flask and count, recording at least 10 000 counts (A^+, A_-). The density of the solution can be taken as 1 g cm⁻³ as it is so dilute. Determine the count corresponding to the total mass of the anode and cathode liquids withdrawn before electrolysis $\frac{1}{2}(w_+A)$ and

† 10 cm³ for a G–M tube.

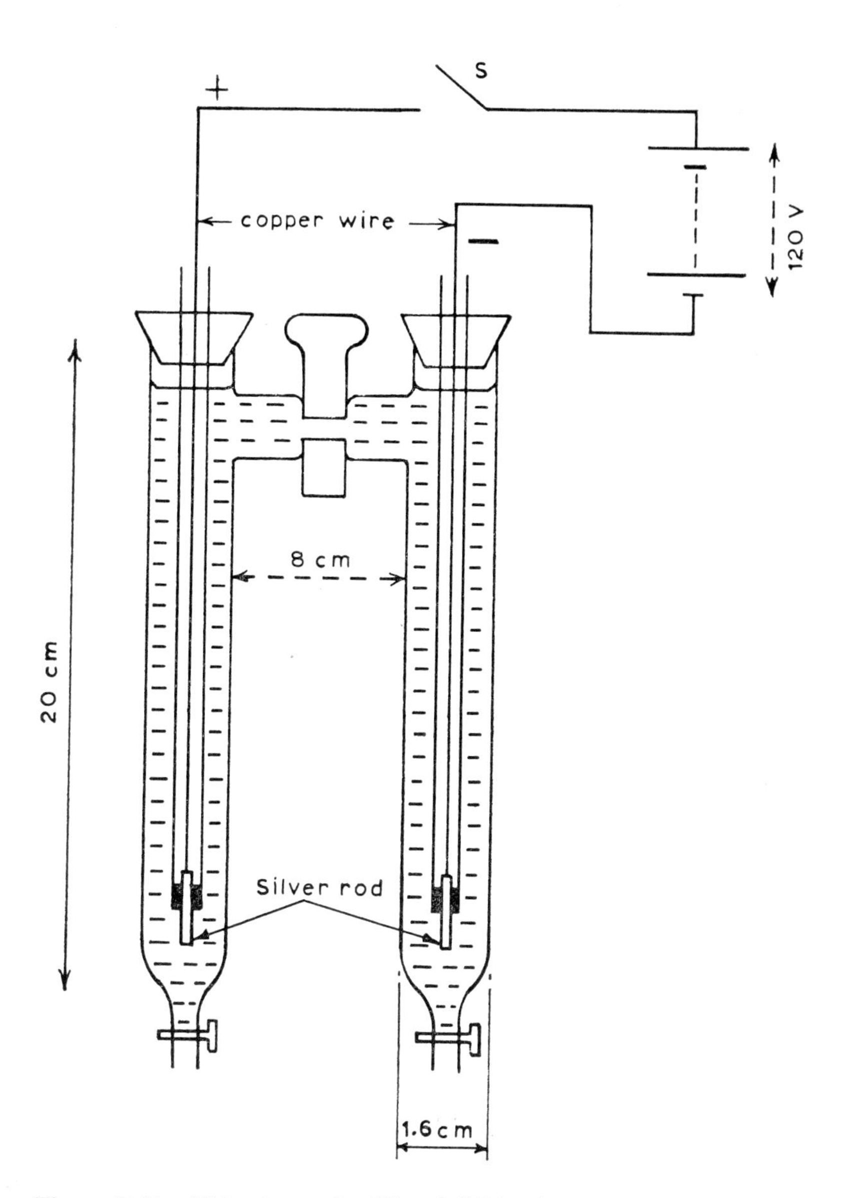

Figure E.22. This shows the Hittorf 'H' tube, used for measuring. transport number. The suggested dimensions give a capacity of 100 cm^3. The silver electrodes are soldered to copper wires and are sealed into the glass tubes with Araldite. The central tap should have a bore of 4–5 mm the corresponding count after electrolysis $\frac{1}{2}(w_+A_+)$, $\frac{1}{2}(w_-A_-)$. The decrease in the anode count$=\frac{1}{2}[w_+(A-A_+)]$, and the decrease in both anode and cathode counts$=\frac{1}{2}[w_+(A-A_+)]+\frac{1}{2}[w_-(A-A_-)]$. Therefore the transport number of the Ag^+

$$=\frac{w_+(A-A_+)}{w_+(A-A_+)+w_-(A-A_-)}$$

Health hazard. All active work must be performed over a tray. There is little health hazard as the quantity of ^{110m}Ag used is 1/240 of the MPAI. The silver nitrate from the Hittorf tube must be poured into the 'Active Waste' bottle. The silver cathode should be dipped in dilute HNO_3 to dissolve the surface layer of active silver deposited. Continue the treatment until all activity has been removed, and pour the nitric acid into the 'Active Waste' bottle.

Experiment 23. To determine phosphate labelled with ^{32}P by radiochemical titration with barium nitrate

Theory. In a radiochemical titration, a substance A in a solution of unknown concentration is labelled with a few μCi of a suitable isotope from a carrier-free solution. A reagent B, which forms a precipitate with A, is chosen. Known volumes of a standard solution of B are added to a series of aliquots taken from the solution of A and the activities of these aliquots are determined with a suitable counter. The activities are plotted graphically against the volume of B added. When the graph reaches a steady horizontal value close to or equal to the background, this is the end-point. The principle is similar to a conductivity titration of a strong acid with a weak base.

As an introduction to the technique, a phosphate has been chosen since there is no method of titrating a phosphate with a normal indicator. The phosphate is precipitated quantitatively in alkaline solution with barium ions.

Apparatus. Scaler/timer with quench unit, liquid counting G–M tube, 2×10 cm^3 burettes, 10 and 5 cm^3 syringe pipettes, centrifuge to take 15 cm^3 tubes, 20×15 cm^3 centrifuge tubes.

Materials. Solution of sodium phosphate, e.g. Na_3PO_4 in approximately 0·1 M ammonia, which is approximately 0·1 M in sodium phosphate (concentration known to the instructor), containing 1 μCi of ^{32}P in 100 cm^3, a solution of barium nitrate 0·25 M, accurately known.

Experimental details. Number the centrifuge tubes 1–20. Into each centrifuge tube add 5 cm^3 of the sodium phosphate solution. This must be done over a tray and the tubes stood in a rack. Add to each tube 0·25, 0·5, 0·75, etc., cm^3 of the barium nitrate solution up to 5 cm^3. Make up each tube to a total volume of 12 cm^3 with water. Centrifuge. While this is proceeding count the background with water in the G–M tube.

Take 5 cm^3 of the sodium phosphate solution, add 7 cm^3 of water, withdraw 10 cm^3 and count for 30 s. This will give the zero reading. Withdraw 10 cm^3 of the supernatant liquid from each tube in turn, starting with No. 1 and count with the G–M tube. 30 s counts should be sufficient for the first few tubes, but the counting time should be increased as the count rate decreases, though it will not be possible to maintain constant statistical accuracy. Do not wash out the G–M tube between counts. Correct all the counts to 1 min counting time, applying a dead time correction to the higher counting rates if necessary, and subtract the background.

Plot a graph of counting rate on the y axis against volume of barium nitrate solution added on the x axis and determine the end point of the titration from the point of intersection of the horizontal portion of the curve with the decreasing count rate. From the graph calculate the concentration of sodium phosphate in mol dm^{-3}.

Health hazard. There is no appreciable health hazard. The precipitates of barium phosphate should be dissolved in dilute nitric acid and poured into the residues bottle. The contents can then be discarded to liquid waste. All the tubes and the liquid counter should be soaked in 5% ‘ Decon ’, as phosphate is difficult to remove from glassware. Gloves should be worn when handling the centrifuge tubes.

Experiment 24. To separate the products of neutron irradiation of a mixture of sodium and potassium chlorides by paper chromatography and autoradiography

(*a*) *To separate the primary activities—Na^+ and K^+.*

Theory. When sodium and potassium chlorides are irradiated with thermal neutrons the following reactions occur:

(1) $^{23}Na+{}^{1}n \rightarrow {}^{24}Na$. $^{24}Na \rightarrow {}^{24}Mg+e+\gamma$ $(t_{1/2}=15$ h).

(2) $^{39}K+{}^{1}n \rightarrow {}^{40}K$. $^{40}K \rightarrow {}^{40}Ca+e+\gamma$ $(t_{1/2}=10^9$ y).

(3) $^{41}K+{}^{1}n \rightarrow {}^{42}K$. $^{42}K \rightarrow {}^{42}Ca+e+\gamma$ $(t_{1/2}=12{\cdot}5$ h).

(4) $^{35}Cl+{}^{1}n \rightarrow {}^{36}Cl$. $^{36}Cl \rightarrow {}^{36}Ar+e$ $(t_{1/2}=4\times 10^5$ y).

(5) $^{37}Cl+{}^{1}n \rightarrow {}^{38}Cl$. $^{38}Cl \rightarrow {}^{38}Ar+e+\gamma$ $(t_{1/2}=37$ min).

The capture cross-sections for ^{23}Na and ^{41}K are 0·54 b and 0·95 b respectively, for thermal neutrons. The abundance of ^{41}K is 6·81%, whereas ^{23}Na has an abundance of 100%. It is necessary to have about eight times as much KCl as NaCl in the mixture. A high specific activity is required. The ^{38}Cl produced will have a relatively high activity at the end of the irradiation.

Apparatus. Scaler/timer with quench unit, G–M tube with mica window of thickness about 2 mg cm^{-2}, rotating drum such as a clynostat (not essential), lead strips to form a slit, strips of Whatman No. 1 chromatography paper about 30 cm length, micro pipette (could be made from capillary tubing), 2 gas jars about 35 cm high with cork and hook attached, strips of X-ray film about 25 × 2·5 cm. Tray with tissues.

Materials. Methanol, concentrated hydrochloric acid, concentrated ammonia, sodium and potassium chlorides (analar grade).

Experimental details. Weigh 5 mg of sodium chloride and 40 mg of potassium chloride into a small plastic stoppered tube and enclose this in a plastic or aluminium outer tube. Irradiate in a reactor with a thermal flux of 10^{12} n cm^{-2} s^{-1} for 1 h. This will produce 35 μCi of ^{24}Na and 31 μCi of ^{42}K, and about 500 μCi of ^{38}Cl. Allow the products to decay for 12 h, or overnight, which will eliminate the ^{38}Cl. Place the plastic tube, using forceps, in a small lead pot and dissolve the chlorides in 0·3 cm^3 of water.

Cut a strip of the chromatography paper about 2·5 cm wide and of such a length that when attached to the hook in the cork of the gas jar the strip is just clear of the bottom. Add 1 cm^3 of the concentrated HCl to 100 cm^3 of methanol and pour the solution into the gas jar to a depth of about 1 cm. Make a pencil line on the paper strip about 2 cm from one end and with the micro pipette transfer about 0·01 cm^3 of the active solution on to the pencil line. The spot must not be larger than 0·5 cm diameter. A little practice may be required before the correct amount is added and the technique should be tried out with water and an odd piece of filter paper. A hair drier is useful to dry out the spot and it is necessary to add the solution in very small quantities to prevent the spot from spreading.

Insert the strip into the jar so that the end is just immersed into the methanol and the pencil line is about 1 cm clear of the liquid. Leave the jar in a position free from draughts until the liquid front has advanced about 25 cm, then withdraw the paper, mark the solvent front with a pencil line and immediately expose the strip to ammonia vapour, which fixes the positions of the various ions. If it is necessary to leave the chromatogram to run overnight a thin line of silicone oil should be painted across the paper just below the top to prevent the solvent front running off the paper. Lastly withdraw the strip and dry with an air blower, or in an oven.

(i) *To prepare an autoradiograph.* When the strip is dry place it in contact with a strip of X-ray film in the dark room. Secure the film with a paper clip, marking the end of the film in contact with the bottom of the paper strip. Leave for several hours, or overnight, and then develop according to the maker's instructions. Three spots should be found, two very dark ones near the bottom and one fainter one considerably higher up. Make a rough measurement of the R_F values (distance moved by the ion divided by the distance moved by the solvent front). The bottom spot is potassium and the second one is sodium.

(ii) *To count a histogram.* Make a second chromatogram in exactly the same way as the first. It can be conveniently started about 20 min after the first if time be available. Dry and fix as before, and attach the strip to a revolving drum such as a clynostat. This should make one revolution in about 30 min. Arrange the G–M tube above the drum so that the paper will pass directly beneath the G–M tube and fix the two strips of lead to make a slit about 0·5 cm wide (see Figure E.24). Set the scaler, and take 15 s counts at 10 s intervals over the full length of the paper. Alternatively the output from the scaler could be fed to a recorder. Note the time taken for the drum to rotate from one pencil mark to the other. This will enable the times to be expressed as R_F values. Plot a graph of counts per 15 s corrected for background, against time and compare the positions of the peaks with the spots of the autoradiograph.

If a rotating drum is not available, the paper strip can be placed under the G–M tube and successive counts taken at 0·5 cm intervals. The procedure will take much longer, but much greater accuracy can be obtained by taking longer counts, e.g. for 1 min. This technique would take at least an hour, whereas with the revolving drum the time taken is of the order of 30 min.

(*b*) *To investigate the secondary activities, detected on the autoradiogram or histogram.*

Allow the ^{24}Na and ^{42}K to decay for 4 days and then repeat the chromatography, but use if possible a thicker paper such as Whatman No. 3MM,

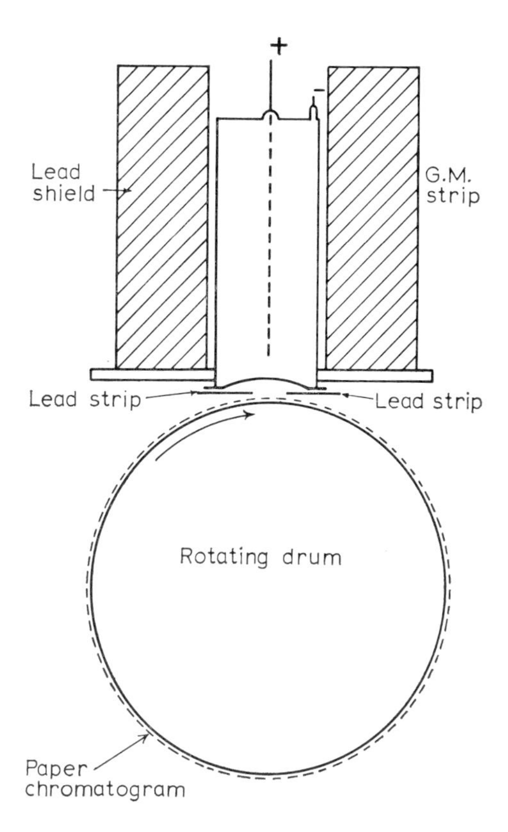

Figure E.24. This is a method of plotting a histogram by using the revolving drum of a clynostat. 15 s counts are taken at 10 s intervals over the whole length of the chromatogram. The time of rotation from the start to the end of the chromatogram must be recorded in order to plot the histogram. Alternatively, the output from the G–M tube could be fed to a recorder.

which would take a loading of 0·04 cm³. It would be advisable to leave the paper in contact with X-ray film for 2 days.

When the autoradiograph is examined it will be found that the two lower spots produced by ^{24}Na and ^{42}K have almost vanished, but the third spot should be much darker. On close examination it may be possible to detect two spots overlapping. It is suggested that these secondary activities should be investigated in two ways:

(1) *By chromatography*. Prepare ion exchange paper by soaking two strips of thick chromatography paper such as Whatman No. 3MM about 25 cm in length, in 20% $Fe(NO_3)_3$ solution. Dry thoroughly in air, and then dip the papers in concentrated ammonia solution. This will precipitate $Fe(OH)_3$ in the paper. Wash the papers until neutral, and dry.

Spot on to each paper about 0·04 cm³ of the active solution and take a count with a thin mica window G–M tube close to the spot. Record the count. Run chromatograms of each paper using 0·05 M $NaHCO_3$ as solvent, in two gas jars. The solvent front should move up about 20 cm in 2 h. Mark the position of the solvent front with a pencil and dry both chromatograms.

Place one of the paper strips in contact with X-ray film. If the activity of the original spot was 20 counts s^{-1} leave the paper in contact with the X-ray film for one week. Make a histogram of the other strip in exactly the same way as with the sodium and potassium. Compare the histogram with the autoradiograph. Two spots should now be found; one at or near the bottom and the other about half way up the paper. (Note that if sufficient activity is not obtained for this investigation the same quantity of sodium and potassium chlorides should be irradiated again in the reactor for a considerably longer time. The chlorides may then have to decay for a week before carrying out the investigation.)

(2) *By plotting a complex absorption curve*. Obtain a dimple planchet and put a tiny smear of silicone oil round the edge. This is to prevent the solution from creeping over the edge of the planchet. Add successive small quantities of the active solution, which has decayed for a week after irradiation, to the planchet and evaporate to dryness under an infrared ray lamp. Continue the additions until a counting rate of at least 3000 counts min^{-1} has been obtained with a mica window G–M tube. Plot a β-particle absorption curve as described in Experiment 10. The curve will show two points of inflection, similar to Figure 9.4. Extrapolate the straight line due to the high energy component backwards to zero thickness, allowing for the thickness of the counter window, and subtract the activity from that of the upper curve due to the two components. Plot the differences and extrapolate the straight line obtained until it meets the base line (see Figure 9.4). From the maximum range of the two components in aluminium determine the maximum β-energy of each component from Figure 9.2. Attempt to identify the two components. It would be possible to determine the half-life of the component with the higher β-energy by making counts at weekly intervals of the spots on the chromatogram prepared in (*a*).

In attempting to determine the nature of the components the student should remember that the reactor spectrum will contain a high proportion of fast neutrons as well as thermal neutrons. He should consider possible fast neutron reactions on the nuclides present (see p. 73).

Health hazard. There is little health hazard as the activities of the two nuclides used for the histogram are 1/50 of the MPAI. The work with the active solution must be performed over a tray.

Experiment 25. To determine the rate constant and activation energy for the exchange reaction between Br^- and bromobutane, in anhydrous ethanol

Introduction. ^{82}Br is a convenient tracer and may be prepared by neutron irradiation of an organic bromide, such as tetrabromoethane, using a laboratory neutron source. (Alternatively ^{82}Br may be purchased.) As explained on p. 153 ^{80}Br and ^{80m}Br are also formed as well as ^{82}Br. If a scintillation counter is used most of the count will come from ^{82}Br. By adjusting the discriminator to cut out the γ-energy below 0·62 MeV the ^{80}Br count could be entirely excluded. With a G–M tube, on the other hand the high energy β-rays from ^{80}Br will produce the principal count.

Theory. The equation for an exchange reaction may be written:

$$RBr + Br^{\circ -} \leftarrow RBr^{\circ} + Br^-$$

where Br° stands for radioactive bromine,
Let $a = [RBr]$, $b = [Br^{\circ -}]$ and $c = [Br^-]$.
Let $x = [RBr^{\circ}]$ at any time t.

In an exchange reaction $K = 1$ and the rate constants for the forward and reverse reactions are equal.

The rate of the forward reaction $= k(a-x)(b-x)$

$$\approx ka(b-x)$$

since x may be neglected in comparison to a.

Similarly the rate of the reverse reaction $= k(c+x)x$

$$\approx kcx$$

since x may be neglected in comparison to c.

The differential equation for the net rate of formation of RBr° (x) is

$$\begin{aligned} dx/dt &= ka(b-x) - kcx \\ &= k\{ab - x(a+c)\} \end{aligned} \qquad \text{(E.25.1)}$$

At equilibrium $dx/dt = 0$
so: $a(b - x_e) = cx_e$
where x_e = the equilibrium concentration of RBr°
hence $ab = x_e(a+c)$
Substituting for ab in Equation (E.25.1),

$$\begin{aligned} dx/dt &= k\{x_e(a+c) - x(a+c)\} \\ &= k(x_e - x)(a+c) \end{aligned}$$

rearranging

$$dx/(x_e - x) = k(a+c)dt$$

integrating

$$-\ln(x_e - x) = (a + c)kt$$

or

$$-\log_{10}(x_e - x) = (a + c)kt/2{\cdot}3 \tag{E.25.2}$$

Since activities are proportional to concentrations provide the same counting equipment is used throughout,

$$\log_{10}(A_e - A) = -(a + c)kt/2{\cdot}3 \tag{E.25.3}$$

By plotting a graph of $\log_{10}(A_e - A)$ against t and knowing the values of a and c, k may be calculated from the slope.

If the experiment is repeated at a different temperature the activation energy (E) may be determined from the equation:

$$\ln k_1/k_2 = -(E/R)(1/T_1 - 1/T_2), \tag{E.25.4}$$

where R is the gas constant (8·32 J mol^{-1}) and T is the temperature in kelvin.

Apparatus. Scaler/timer with well-type scintillation counter, or liquid counting G–M tube in castle with quench unit, special apparatus (Figure E.25), neutron source or supply of ^{82}Br (2 μCi would be sufficient), 300 cm^3 stoppered separating funnel, small stoppered separating funnel, 2×10 cm^3 and 2×5 cm^3 measuring cylinders, 1 and 2 cm^3 pipettes with syringes, hot plate, supply of specimen tubes to fit the well of the scintillation counter. They should hold 4 cm^3. (If a well-type scintillation counter is not available use 50 cm^3 beakers and stand on top of the counter). 2 large trays with tissues.

Materials. 0·4 M solution of 2-methyl-2-bromopropane (tertiary butyl bromide) in anhydrous ethanol (10 cm^3 required), 1·0 M solution of sodium bromide in water (2 cm^3 required), tetrabromoethane (either symmetrical or unsymmetrical) 200 cm^3 approximately, which contains about 3×10^{-4} mol of bromine. (This solution must have been irradiated with the neutron source for about 10 days before the experiment). Anhydrous ethanol 20 cm^3, methanol 400 cm^3, cyclohexane 50 cm^3, 20% solution of sodium nitrate 100 cm^3.

Experimental details. Before using the radioactive tracer the student should familiarize himself with the technique for extracting liquid from the inner tube A (see Figure E.25). A rubber bulb, or a supply of compressed air or nitrogen, should be connected to C. The tube F should be closed with a finger and the application of gentle pressure at C will push the required volume of liquid from A into the measuring cylinder E.

To carry out the experiment stand the whole apparatus on a large tray and cover the bench with ' Benchcote ' or other absorbent paper. Pour 10 cm^3 of the 2-methyl-2-bromopropane solution into the tube A and fill the 3-necked flask B three-quarters full with methanol. The volumes quoted are suitable for an inner tube of capacity 25 cm^3. Switch on the mantle and boil the methanol gently under reflux.

While the apparatus is reaching temperature equilibrium, take 2 cm^3 of the sodium bromide solution, measured by pipette and make up to 10 cm^3 approximately with water. Extract the radioactive bromine from the tetrabromoethane, which has been irradiated with the neutron source, by shaking it with the sodium bromide solution for at least 5 min. Thorough agitation is essential. Under these conditions almost complete exchange takes place between the free radioactive bromine in the tetrabromoethane and the bromide ions in the aqueous layer. (See Chapter 10 (p. 143) for the theory of this exchange reaction.) Allow the two layers to separate, run off the lower layer of tetrabromoethane into a bottle, and the upper aqueous layer into a small beaker. These operations should be performed over a tray.

Evaporate the aqueous solution in the beaker to dryness and test the sodium bromide residue with a monitor. It should be strongly active. Redissolve the sodium bromide in 10 cm³ of anhydrous ethanol. It will dissolve slowly.

Pipette 2 cm³ of this solution into a specimen tube and count for 100 s with the scintillation counter. Return the 2 cm³ as completely as possible

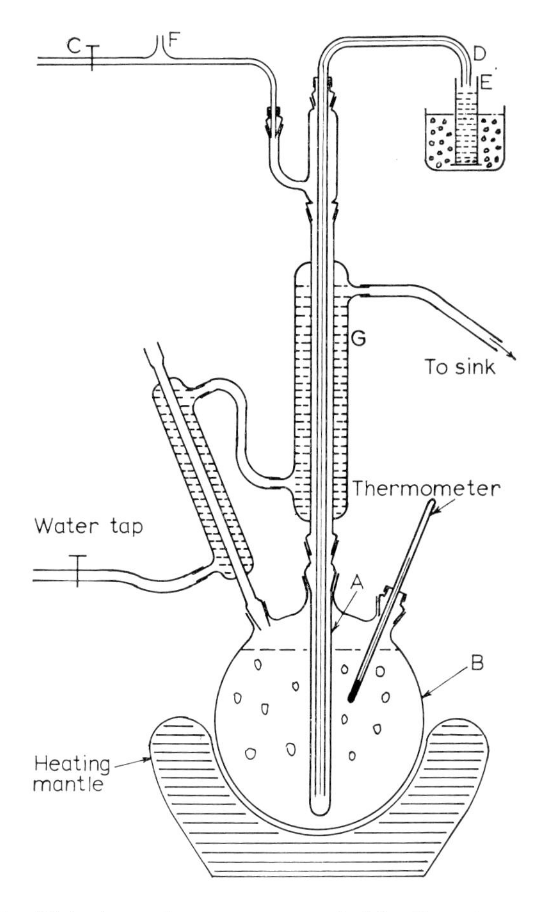

Figure E.25. This shows the apparatus needed for Experiments 25 and 26. The flask B should have a capacity of 500 cm³. A convenient size for A, which contains the reaction mixture, is 20 or 25 cm³. E is a small measuring cylinder (5 cm³). A supply of compressed air should be attached to C or the tap may be replaced by a rubber bulb. This enables portions of the reaction mixture in A to be withdrawn and collected in E at known time intervals. Suitable liquids for use in B are methanol and ethanol.

to the remainder of the solution. Count 2 cm^3 of the tetrabromoethane to calculate the efficiency of extraction. This should be at least 45%.

Now transfer the 10 cm^3 of alcoholic sodium bromide solution to a small flask and heat it in hot water to the temperature of the boiling methanol. Transfer the solution as quickly as possible into the tube A and stir by blowing compressed air gently through the mixture. Record the time and take this as zero time.

Surround the measuring cylinder E with ice and water and blow about 1·5 cm^3 of the contents of the tube into the measuring cylinder. Take 1 cm^3 by pipette, transfer to a small specimen tube, and count with the scintillation counter with the discriminator set at 0·66 MeV integral. Mark this tube S, record the time at half count, and set the tube aside for counting at intervals. The contents should decay with $t_{1/2}=36$ h if the discriminator has been correctly set. Otherwise the graph of log activity of S against time will be a curve due to counts from ^{80}Br, but this graph can be used to correct all count rates to zero time.

After about 5 min blow about 2·5 cm^3 of the contents of the reaction mixture into the measuring cylinder, recording the time from zero. Pipette 2 cm^3 into a small stoppered separator, add about 5 cm^3 of 20% sodium nitrate solution followed by 4 cm^3 of cyclohexane. Shake vigorously. The bromobutane will be extracted quantitatively into the cyclohexane leaving the bromide ions in aqueous solution. Allow the two layers to separate and run the lower aqueous layer into a flask labelled ‘ radioactive waste ’; run the cyclohexane into a specimen tube which must fit inside the well of the counter, or use a small beaker to stand on top of the counter. It is most important to achieve a clean separation from the aqueous layer as in the early stages the aqueous layer will be much the more active of the two. With the discriminator set as above, count the background, if this has not already been done, and then count the cyclohexane specimen for a suitable time, e.g. 1000 s, recording the time half-way through the count.

Repeat the above mentioned procedure at 25, 55, 95, 160 and 240 min from zero time. Finally withdraw the inner tube and heat the contents under reflux for a further 30 min to bring the reaction to equilibrium. Withdraw a final portion and count as before.

Using the log graph obtained with tube S, correct all the counts to zero time. The final reading will be the equilibrium value (A_e)†.

Tabulate the results as follows:

No.	Time of withdrawal (t)	Time of count (t_1)	Counts per 1000 s	Counts − background	Counts corrected for decay to $t_1=0$	A_e-A	$\log_{10}(A_e-A)$

Plot a graph of $\log_{10}(A_e-A)$ against t. k may be determined from the slope of the graph and the known values of a and c.

† The equilibrium value A_e, should be 0·5 of the total activity present. In practice the author has found, after repeated experiments, that A_e is much lower than this, owing to side reactions. Nevertheless, a good straight line of $\log_{10}(A_e-A)$ has invariably been obtained.

Repeat the experiment using ethanol in the flask B instead of methanol. The reaction will proceed much more rapidly than with methanol as the heating liquid, and portions should be withdrawn at intervals of 15, 35, 60, 90, 125 and 160 min from zero time.

From the two values of k determine the activation energy using Equation (E.25.4).

Note on the activity to be expected. The irradiation of 200 cm^3 of tetrabromoethane to saturation with a neutron source giving a flux of 10^3 n cm^{-2} s^{-1} would produce a total activity of 0·2 μCi of ^{82}Br. Assuming 50% extraction, one-tenth of the total activity withdrawn each time from the reaction tube and a maximum of half of this counted at equilibrium, a final count corrected for decay should be about 20 counts s^{-1}. The initial counting rate will be much less. Counts for 1000 s would be advisable.

Modifications when using a G–M tube. The principal count will come from the ^{80}Br. As both ^{80}Br and ^{80m}Br are formed independently by neutron irradiation (see p. 153) it will simplify the measurements if the independent ^{80}Br is allowed to decay. The first count should be made about 1·5 h after the initial extraction from the tetrabromoethane. Equilibrium will then have been reached between the two nuclear isomers and the count will be caused by the 2 MeV β-particles from ^{80}Br. The correction of the counts made at different times should be made in the same way as previously described. Use 10 cm^3 of cyclohexane for the extraction.

Modifications when using ^{82}Br prepared by reactor irradiation. 0·025 g of NH_4Br should be irradiated in a flux of 10^{12} n cm^{-2} s^{-1} for 1 h. 20 μCi of ^{82}Br will be produced together with more than 1 mCi of ^{80}Br and ^{80m}Br. It would be advisable to allow the NH_4Br to decay for 48 h to remove the activity of the ^{80}Br and ^{80m}Br. This would leave about 8 μCi of ^{82}Br, which would be sufficient for four experiments.

Dissolve the irradiated NH_4Br in 8 cm^3 of the 1·0 M NaBr solution. Take 2 cm^3 of this solution by pipette, evaporate to dryness and redissolve in 10 cm^3 of anhydrous ethanol. Bring the solution to the temperature of the bath and add as before to the tube A. Proceed as previously explained.

Health hazard. There is no appreciable health hazard as even with the reactor-irradiated bromide the amount of activity used by the student is 1/1000 of the MPAI, but all operations with active material should be done over a tray. The handling of the neutron source and reactor-irradiated ammonium bromide during the first 48 h must be done by a qualified instructor or technician.

Experiment 26. To determine the rate constant and activation energy for the exchange reaction between I^- and iodoethane, in anhydrous ethanol

Theory. The student is referred to Experiment 25 for the theory of exchange rate. The reaction is carried out as before in anhydrous ethanol using the apparatus shown in Figure E.25. Pentane (boiling point 39°C) is used instead of methanol. At this temperature the experiment can be completed in 1 h. If a neutron source is available ^{128}I can be used, otherwise it is necessary to use ^{131}I.

Apparatus. As for Experiment 25. Liquid counting G–M tube with quench unit required for ^{128}I, but well-type scintillation counter preferable for ^{131}I; tray with tissues.

Materials. Anhydrous ethanol 50 cm³, pentane 400 cm³, iodoethane 200 cm³ (to prepare ^{128}I) cyclohexane 50 cm³, 20% solution of $NaNO_3$ 100 cm³, NaI.

Experimental details. Dissolve 0·06 g of iodine in approximately 200 cm³ of iodoethane and irradiate with a neutron source for 2 h. If the source yields $< 10^6$ n cm^{-2} s^{-1}, it is better to immerse the source (in a plastic tube) in the iodoethane. This will give the highest yield of ^{128}I.

Set up the apparatus as shown in Figure E.25 with pentane gently boiling. Dissolve 1 cm³ of iodoethane (measured by pipette) in a total volume of 10 cm³ of anhydrous ethanol and pour the solution down the condenser into the tube A.

Dissolve 1·800 g of NaI in about 10 cm³ of water and pour into the large separator. Wash the vessel containing the sodium iodide with a few drops of water and add the washings to the separator. Withdraw the 200 cm³ of iodoethane from the neutron source and without delay pour into the separator. Shake the contents vigorously for 2–3 min. Allow the two layers to separate, run the iodoethane into a flask, separating it as completely as possible from the aqueous layer and run the aqueous layer into a 100 cm³ beaker, washing the separator with a few drops of water into the beaker. Evaporate the solution of labelled NaI on a hot plate, or over a small bunsen to dryness, and redissolve in 10 cm³ of anhydrous ethanol. The whole operation of extracting the ^{128}I into the NaI, evaporation to dryness and resolution in anhydrous ethanol should not take more than 25 min, i.e. one half-life of ^{128}I. Using a 0·5 Ci Am–Be neutron source and irradiating 200 cm³ of iodoethane the author obtained a count rate of the aqueous layer, corrected for dead time, of 60 000 counts min^{-1}. The efficiency of extraction was 66%.

Warm the alcoholic solution of NaI to 40°C and pour it into the tube A, starting a stop-clock. Agitate the contents of A by gently blowing air through D into the solution, with a squeeze bulb.

After 2 min withdraw about 2·5 cm³ of the reaction mixture, and using the technique described in Experiment 25, p. 218, extract 2 cm³ with 10 cm³ of cyclohexane; the cyclohexane extracts the iodoethane quantitatively. Separate the cyclohexane layer from the aqueous layer and run the cyclohexane layer into the liquid counting G–M tube. Count for 300 s, noting the time on the stop-clock half-way through the count.

Repeat this whole procedure at 5 min, 10 min, 20 min, 40 min, and 60 min. Finally, withdraw the inner tube with the condenser, immerse the tube in a beaker of water, heated to 80°C for 5 min to complete the reaction, withdraw a specimen, extract and count as before.

Plot a graph of the decay of ^{128}I on semi-log paper, against time, remembering the half-life of ^{128}I is 25 min, and use this graph to correct all counts to zero time. Tabulate the results as in Experiment 25, plot a graph of $\log_{10}(A_e - A)$ against t and determine the rate constant. The values of a and c (Equation E.25.3) can be determined from the known concentrations of iodoethane and NaI. The density of iodoethane is 1·933 g cm^{-3}.

Repeat the experiment using 2,2-dimethylbutane (b.p. 49°C) as the boiling liquid. The specimens should be withdrawn at more frequent intervals as the reaction should be almost complete in 45 min. Use the two values of k to calculate the activation energy (Equation E.25.4).

Health hazard. There is no health hazard as the initial quantity of ^{128}I produced is about 1 μCi and the half-life is so short. The irradiation with the neutron source must be performed by a qualified instructor, or technician.

Modification using ^{131}I. In this case dissolve 1·800 g of NaI in 10 cm^3 of anhydrous ethanol and add 1 μCi of ^{131}I from a concentrated stock solution. If the quantity of aqueous solution of ^{131}I required does not exceed 0·01 cm^3 the amount of water added to the reaction mixture will not cause appreciable hydrolysis of the iodoethane. A well-type scintillation counter should be used and the volume of cyclohexane used to extract the active iodoethane should be adjusted to suit the depth of the well.

Health hazard. ^{131}I is an isotope of medium toxicity. Rubber or polythene gloves should be used when shaking the separating funnel.

Experiment 27. To separate ^{239}Np from uranium by ion exchange

(*a*) *After neutron irradiation of uranium with a reactor.*

Theory. Neptunium-239 can be prepared by irradiating a uranium compound in the neutron flux of a reactor. ^{239}U is first formed and decays ($t_{1/2}$=23·5 min) to ^{239}Np. The neptunium may be separated using a cation exchange column in the hydrogen form and eluting with hydrochloric acid. Identification is best carried out by a half-life measurement (2·3 d), using either a γ-spectrometer and counting on the 0·106 MeV peak, or by β-counting with a liquid counting G–M tube. Identification of the 0·106 MeV peak is not sufficient as ^{234}Th has a peak at 0·10 MeV.

Apparatus. Scaler/timer, with single channel pulse height analyser and scintillation counter, or liquid counting G–M tube, glass tube 30 × 0·4 cm approximately, fitted with a tap and funnel for filling, dropping tube to read to 0·1 cm^3, 20 specimen tubes, capacity 2–3 cm^3, tray with tissues, block of expanded polystyrene with 20 holes to take the specimen tubes.

Materials. Cation exchange resin, either Zeo-carb 225 or Dowex 50, uranyl(VI) nitrate, 3·2 M hydrochloric acid.

Experimental details. Weigh a small quantity of uranyl(VI) nitrate into a transparent polyester or polythene stoppered specimen tube and dissolve the uranyl(VI) nitrate in 1 cm^3 of molar sulphuric acid. With a neutron flux of 10^{12} n cm^{-2} s^{-1} and 1 h irradiation, 0·02 g of uranyl(VI) nitrate will yield about 20 μCi of neptunium and a variable quantity of fission products, depending on the degree of depletion of the ^{235}U. Most uranium salts now contain about 0·4% of ^{235}U. 20 μCi of neptunium would be sufficient for about 10 experiments if used within two days of the irradiation.

Place the specimen tube, containing the uranyl(VI) nitrate, with some polystyrene packing in a 7·5 cm by 2·5 cm standard can, or seal it in a polythene tube according to the reactor requirements. Irradiate in the reactor for a suitable time. Allow 4 h to elapse after the irradiation so that the ^{239}U will have effectively decayed to ^{239}Np.

Fill the glass tube with the resin to a depth of about 20 cm and condition it to 3·2 M hydrochloric acid, by running about 3 bed volumes of the acid slowly through the resin. The effluent must be checked by titration. Drain to bed level. Mark the specimen tubes 1–20 and place them in the holes in the polystyrene block for convenience. Add 0·1 cm³ of the irradiated uranyl(VI) nitrate solution on to the top of the column and again drain to bed level. Fill the remainder of the tube and the funnel with 3·2 M hydrochloric acid and run 1 cm³ (about 20 drops) into the first specimen tube at a rate of about 1 drop every 2 s. Repeat with each tube in turn.

Adjust the scaler/timer and pulse height analyser to 0·106 MeV and set the gate width to 5% or nearest. Count the background. It would be convenient to do this while the elution is taking place. Measure the activity of each tube. A counting time of 30 s should give sufficient accuracy, but this depends on the sensitivity of the apparatus. Plot a graph of counting rate, corrected for background against the volume of eluant. The neptunium should be eluted in the first five tubes as it will be reduced to Np(V) by the resin (p. 52). As Np(V) exists as NpO_2^+ it is the most easily eluted oxidation state. The uranyl(VI) ions, fission products and ^{234}Th are all held on the resin, under these conditions.

Select the tube with the highest counting rate and count at least 10 000 counts. Record the time and date. Repeat at daily intervals, or more frequently, for a week, and determine the half-life (see Experiment 4, p. 174).

If β-counting is used the contents of each specimen tube should be made up to a particular volume to suit the capacity of the G–M tube.

Health hazard. There is no appreciable health hazard in handling the 2–3 μCi of neptunium used in the experiment. The irradiated uranium solution must be kept in a suitable lead pot and all operations conducted over the tray. The resin in the tube should be put in a polythene bag and kept to decay until it can be disposed of as solid waste, or the fission products may be eluted with concentrated hydrochloric acid and disposed of as liquid waste along with all the solutions. Take particular care to wipe the specimen tubes before placing on the scintillation counter. Read Chapter 12 before doing the experiment.

(*b*) *Separation after neutron irradiation of uranium with a neutron source.*

Theory. The experiment just described is easy to carry out. It is much more difficult, but also more instructive, to perform the preparation with a neutron source. In this case it is necessary to irradiate the uranium until the maximum amount of neptunium has been produced, which will be after five or six half-lives. (See Chapter 5, p. 69, and Equation (5.15).)

Apparatus. Scaler/timer with pulse height analyser and scintillation counter (preferably high resolution), glass tube 1·5 m long by about 2·5 cm diameter, fitted with a rubber cork and tap, or glass tube and rubber tube with screw clip, a large number of 25 cm^3 beakers, 100 or 250 cm^3 separating funnel with rubber cork to fit the glass tube, nitrogen or air cylinder, tall retort stand or scaffolding, two large trays with tissues, neutron source with an output of at least 10^6 n cm^{-2} s^{-1}.

Materials. Cation exchange resin, either Zeo-Carb 225 or Dowex 50, 100–200 mesh—500 g, uranyl(VI) nitrate (preferably depleted in ^{235}U), 100 g, 3·2 M HCl, 1000 cm^3.

Experimental details. Dissolve 100 g of $UO_2(NO_3)_2.6H_2O$ in the minimum of 3·2 M HCl (about 75–100 cm^3) in a 250 cm^3 plastic beaker. Stand the beaker on top of the scintillation counter and set the pulse height analyser to 0·106 MeV with a 5 or 10% gate width. Count for 1000 s and record the nett count after subtracting the background.

Fill a plastic bucket with water and clamp the beaker, containing the uranium solution, so that it is immersed in the water to within 4·5 cm of the lip. This must be set up in a room or cupboard which is apart from the laboratory and which can be locked. Fix a plastic tube, wide enough to take the neutron source, in the middle of the beaker. Withdraw the neutron source from its housing, using a long rod which can be screwed into the source, or long-handled tongs, and insert it into the plastic tube. The level of the uranium solution should be at least 2 cm above the top of the neutron source. If necessary add more 3·2 M HCl. Leave this set up for 10–14 days. On no account use borosilicate glass instead of plastic or the flux will be reduced considerably by neutron capture by the boron.

Put a wad of glass wool into the 1·5 m tube above the cork and add some water. Pack down the glass wool to make a plug about 1 cm thick to hold the resin. Drop the resin slowly into the water, or slurry it in mixed with water. Make sure no air is trapped in the resin; stirring with a long metal rod is helpful. When all the resin has been added drain the water to bed level. Apply a pressure of about 30 kN m^{-2} (5 lb in^{-2}).

Withdraw the neutron source from the uranium solution after the appropriate time and, placing the beaker with its contents on top of the scintillation counter, count again for 1000 s, after allowing about 2 h for the residual ^{239}U to decay. Record the nett count with the same setting of the pulse height analyser and compare to the previous count. The difference will be a rough measure of the activity of neptunium produced. Unless the second count is at least 2000 counts per 1000 s higher than the first there is no point in proceeding further.

A quick check on the yield of Np which can be obtained with the neutron source available may be carried out by determining the yield of ^{239}U. In this case count the uranium solution before irradiation for 300 s, with the pulse height analyser set to 0·074 MeV and 10% gate width, since this is the energy of the principal γ-ray produced by ^{239}U. Withdraw the neutron source from the beaker after a minimum of 3 h. Note the time. Count the contents of the beaker with the pulse height analyser set as above. Subtract the previous reading and correct the difference for decay.

The relative efficiencies of detection for ^{239}U and ^{239}Np are approximately in the ratio of $84/18 \approx 5$ since 84% of the ^{239}U atoms decay with the emission of a 0·074 MeV γ-ray, whereas only 18% of the ^{239}Np atoms emit the 0·106 MeV γ-ray. The maximum count obtained from the ^{239}Np on withdrawal of the neutron source could only be about a fifth of that obtained with the ^{239}U. This procedure will give a good indication of the yield of ^{239}Np which could be obtained with the neutron source available. If the calculated yield is less than 600 counts per 300 s a different arrangement of the neutron source and target should be tried.

It is of the utmost importance to have the ^{239}Np as Np(V) before attempting the separation. In the author's experience ^{239}Np is produced as Np(IV) when a large excess of uranyl(VI) nitrate is irradiated in aqueous solution with a neutron source. In order to be sure that the ^{239}Np is Np(V) first oxidize to Np(VI). The uranium solution should be made 1·0 M in HNO_3 and $KMnO_4$ solution added until a permanent pink colour is obtained. This procedure will rapidly oxidize the Np(IV) to Np(VI). Allow 15 min at 25°C for complete oxidation. A few drops of hydrazine solution should now be added. This will rapidly reduce the Mn(VII) to Mn(II) and the Np(VI) to Np(V). Do not use excess hydrazine as hydrazine slowly reduces Np(V) to Np(IV). Now make the solution 3·2 M in HCl, add one further drop of the hydrazine solution and pour the uranium solution on to the resin running it very slowly through the column. Do not exceed a flow rate of 1 cm^3 min^{-1}. Apply pressure if necessary. Collect the effluent in 5 cm^3 quantities in the small beakers, which should be numbered serially. When the uranium solution has fallen to bed level run on 3·2 M HCl. The volume of liquid in the resin bed will be about 150 cm^3 so that neptunium is unlikely to appear until at least this volume has run through. However, each beaker should be counted for 300 s with the above setting of the pulse height analyser.

Plot the nett counts against volume of effluent. When the first peak has been reached collect all the solution which corresponds to this peak, concentrate to a small volume and count it at daily intervals. Plot a log graph of the count rate against time in hours and determine the half-life. If the half-life is incorrect the elution has been too fast and the whole experiment must be repeated. (A small peak caused by anionic fission products may appear before the ^{239}Np.)

To clear the resin of uranium, thorium and fission products run through a saturated solution of ethanedioic acid (oxalic acid) until all activity has been removed. An end-window G–M tube placed close to the column will record the movement of the ^{234}Th through its daughter ^{234}Pa. ^{234}Th will be the last nuclide to leave the resin. Ethanedioic acid removes the uranium and thorium by complexing.

Health hazard. There is no health hazard in this experiment, but the concentrated solution of uranium must be handled over a tray and a deep tray or trough must be placed under the column to guard against a spill. All operations with the neutron source must be carried out by a qualified instructor or technician.

Experiment 28. To prepare benzenecarboxylic acid (benzoic acid) labelled with ^{14}C in the carboxyl group

Theory. This experiment gives a student an introduction to the handling of hazardous substances under vacuum. It illustrates an important method of labelling organic compounds with ^{14}C. The method can be applied directly to the preparation of any carboxylic acid and by subsequent reduction a variety of compounds can be prepared.

Labelled carbon dioxide, generated by the action of concentrated sulphuric acid on labelled sodium carbonate is passed into a cooled dilute solution of a Grignard reagent, in this case phenylmagnesium bromide. The resulting complex is decomposed with acid, the benzenecarboxylic acid extracted from the ether with sodium hydroxide and precipitated by acidification. It is then recrystallized.

Apparatus. Vacuum line (Figure E.28 (*a*)), special filter flask (optional Figure E.28(*b*)), insulated glass dish to surround flask H, vacuum flasks, nitrogen cylinder, 3-necked flask to prepare Grignard reagent (fitted with tap funnel and condenser), 100 cm^3 separating funnel, tray with tissues.

Materials. Grignard magnesium turnings, sodium-dried ether, dry bromobenzene, concentrated sulphuric acid, sodium carbonate labelled with ^{14}C (50–100 μCi), sodium carbonate carrier, liquid nitrogen, methylbenzene.

Experimental details. A simple vacuum line is illustrated in Figure E.28 (*a*). A is a glass stopper (a Pirani gauge could be inserted here). B are taps preferably of oblique bore. B_6 is a wide oblique bore tap of 3–4 mm. C is a CO_2 generator. D is a tube containing molecular sieves or ‘anhydrone’ as drying agent. E is a 1000 cm^3 flask with ‘prong’ to collect the $^{14}CO_2$ (the volume of this flask should be known). F is a mercury manometer. G is filled as for D. H is a 3-necked flask containing the Grignard reagent. I is a motor, preferably brushless, attached to a stirrer made of piano wire, well greased, which passes through the rubber cork: this will hold the vacuum for one experiment. L is a nitrogen inlet.

Before starting the experiment make sure that the apparatus has no leaks. Evacuate, close tap B_6 and observe the manometer or Pirani gauge over several hours. All taps and ground glass joints must be lightly smeared with a good quality vacuum grease (not silicone). Do NOT attempt to turn the taps unless they have been warmed first with a hot air blower.

Prepare the necessary quantity of $^{14}CO_2$. Take 1·06 g of Na_2CO_3 and add 50–100 μCi of ^{14}C as $Na_2{}^{14}CO_3$. Heat the solid to 250°C for one hour, cool in a desiccator and transfer to the flask of the CO_2 generator. Add 2·5 cm^3 of fresh concentrated sulphuric acid to the tap funnel of the generator and connect to the vacuum line. With all taps closed except B_3, B_6 and B_7, pump out E. Close B_3, open B_2 and pump out the CO_2 generator. Close B_6 and record the reading of the manometer. Open B_3 and very cautiously open B_5 allowing sulphuric acid to react with the sodium carbonate. When all the sulphuric acid has been added (there is a large excess) warm the flask gently to complete the reaction. Immerse the ‘prong’ of E in liquid nitrogen to solidify the CO_2 and when the manometer has returned to its original reading close B_2. Remove the liquid nitrogen from E and when the apparatus had warmed up to room temperature take the reading on the manometer. From the known volume of E and the remainder of the vacuum line, and the pressure of the gas, determine the number of moles of CO_2 collected. If the reaction in C has gone to completion there should be 10 mmol. Cool E once more with liquid nitrogen and when the manometer reads zero pressure close B_3. The experiment may be interrupted at this stage.

Prepare the Grignard reagent under nitrogen; the presence of oxygen will produce carbinol by-products which are difficult to remove from benzenecarboxylic acid. Add 30 cm^3 of dry ether and 0·65 g of Mg turnings to a 3-necked flask fitted with a tap funnel, condenser and inlet tube dipping into the ether. Flush the flask slowly with dry nitrogen. Reduce the flow of nitrogen and add 4 g of dry bromobenzene dissolved in 20 cm^3 of dry ether to the tap funnel. Add about one-tenth of the bromobenzene solution to the flask and warm gently with warm water. If the reaction does not start add a crystal of iodine; Grignard reactions are often very slow to start. As soon as the reaction starts add slowly the remainder of the bromobenzene solution so that the ether continues to boil. When the reaction is complete assay the Grignard reagent by withdrawing 5 cm^3 of the ether solution with a

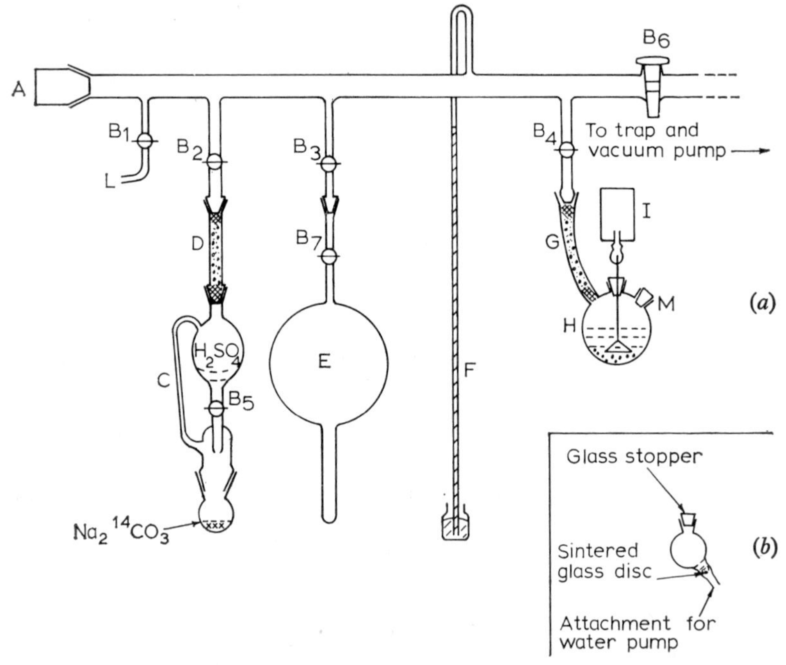

Figure E.28. (*a*) This shows the vacuum line suitable for Experiment 28. The main tube is 2 cm in diameter and of length about 70 cm. B_1 is for admission of nitrogen. D and G both contain drying agents. The volume of the upper bulb in C is about 80 cm^3 and the lower about 100 cm^3 or less. The flask E has a volume of about 1000 cm^3 (known) and H of 250 cm^3. I should be a brushless motor. The tube F must be 90 cm long. CO_2 labelled with ^{14}C is generated in C and collected in E by immersing the prong in liquid nitrogen. The stirrer in H is made of piano wire and passes through the cork, lubricated with grease. H may be surrounded with a bath of solid CO_2/acetone, or liquid nitrogen/toluene at $-20°C$ to draw in the CO_2 from E, or with liquid nitrogen for the purpose of evacuation. (*b*) The flask has a volume of 100 cm^3 and is fitted with a glass stopper and sintered glass disc. It enables active precipitates, such as benzenecarboxylic acid labelled with ^{14}C, to be collected and recrystallized *in situ*.

pipette, which has been flushed with nitrogen and is fitted with a syringe. Discharge the contents of the pipette into 50 cm³ of standard 0·1 M HCl which should be cooled in ice. Titrate the excess HCl with standard 0·1 M NaOH. Calculate the number of mols of Grignard reagent which have been produced.

Now add rapidly to H sufficient of the Grignard reagent to have a slight excess with regard to the CO_2 in E. If E contains 10 mmol of CO_2 add 12·5 mmol of Grignard reagent and make the volume up to 50 cm³ with dry ether. The concentration of the Grignard reagent should not exceed 0·25 M or precipitation may occur on cooling. Flush flask H with nitrogen gently by connecting the cylinder to L, opening taps B_1 and B_4 and removing the stopper M. Close B_1, quickly replace the stopper M and freeze the ether solution by surrounding with liquid nitrogen. With B_4 open pump out all the nitrogen. Close B_4 and surround H with a bath at -20°C which can be made by mixing liquid nitrogen with methylbenzene, or by using dry ice and propanone. This will allow dissolved nitrogen to escape from the Grignard solution and aid carbonation. Stirring will help. Freeze H once more and pump out the nitrogen which has escaped.

Warm up the Grignard reagent to -20°C again, and with all taps closed except B_4 start the stirrer and slowly open B_3. The height of the manometer should not fall by more than 50 cm during the carbonation which should be complete in 15–20 min. To draw in the last traces of CO_2 the contents of H should again be frozen with liquid nitrogen. When the pressure falls to zero close B_4, warm H to -20°C and stir for a further 15 min. Pass in nitrogen through B_1 and B_4 and vent the flask H to the atmosphere. The experiment may be interrupted at this stage (replace stopper M).

Decompose the Grignard reagent complex by adding dilute hydrochloric acid and cooling the flask in ice water. Now, working over a tray, transfer the contents of the flask to a 100 cm³ separating funnel, separate the ether layer (upper) and extract the aqueous layer with two lots of 25 cm³ of ether. Discard the aqueous layer, return the three ether lots to the separating funnel and extract twice with 10 cm³ of 2 M NaOH solution. This process separates the benzenecarboxylic acid from by-products such as ketones, diphenyl and benzene.

Acidify the sodium hydroxide solution with 10 cm³ of concentrated HCl and cool in ice. This precipitates the benzenecarboxylic acid. The process is conveniently carried out in the flask with attached sinter (Figure E.28 (*b*)), and recrystallization and collection can conveniently be carried out in the apparatus. The benzenecarboxylic acid is collected on the sinter and can be detached by gently blowing with a rubber bulb. Alternatively the benzenecarboxylic acid may be collected on a small Hirsch funnel or sinter. The HCl solution should be extracted with ether to collect traces of benzenecarboxylic acid and the ether solution extracted with NaOH solution. On acidification of the NaOH a further small crop of crystals will be obtained. Both crops of crystals should be collected together and recrystallized from water, using the smallest possible volume, and cooling in ice. The yield should be 80% of theory.

Health hazard. There is no appreciable health hazard working with this quantity of ^{14}C, but it would be a good idea to have the vacuum line in a fume cupboard, or under a hood. All apparatus used should be well rinsed with sodium carbonate. The operations with the solid $Na_2{}^{14}CO_3$ should be carried out over a tray.

Experiment 29. To determine the molecular complexity of ^{14}C-labelled benzenecarboxylic acid (benzoic acid) in methylbenzene (toluene)

Theory. The radiochemical technique about to be described avoids the problems associated with the use of very dilute barium hydroxide and CO_2-free water, both of which are essential for the classical technique. Before doing this experiment the student must be acquainted with the technique of using liquid scintillation counting (Experiment 12). For the theory of liquid scintillation counting see Chapter 7, p. 111.

The distribution law states that when a substance is distributed between two immiscible solvents, the ratio of the concentrations is a constant provided no association or dissociation occurs; i.e. if X stands for the substance and a and b for the two solvents:

$$X_a \rightleftharpoons X_b$$

and if C stands for the concentration of X in g dm^{-3}

$$\frac{C_b}{C_a} = k_1 \tag{E.29.1}$$

If, however, association or dissociation occurs the above simple law does not hold. Consider a case where X exists mainly as single molecules in a and mainly as X_n in b. Then in addition to Equation (E.29.1) which applies only to the equilibrium between the single molecules in each solvent, there exists the equilibrium between the single and associated molecules in b, i.e.

$$nX_b \rightleftharpoons X_{n,b}$$

and, by the equilibrium law,

$$\frac{C_{n,b}}{C_b{}^n} = k_2 \tag{E.29.2}$$

Raising Equation (E.29.1) to the nth power and substituting for $C_b{}^n$ in (E.29.2) gives:

$$\frac{C_{n,b}}{C_a{}^n} = k_2 . k_1{}^n = k_3 \tag{E.29.3}$$

And, taking logs:

$$\log C_{n,b} = n \log C_a + \log k_3$$

The slope of the graph will give the value of n, and the intercept the value of $\log k_3$.

Apparatus. Liquid scintillation counting head with scaler/timer, 9 liquid scintillator phials, thermostat set for 25°C, 4×15 cm^3 centrifuge tubes with well fitting bark corks, 2×1 cm^3 syringe pipettes, 50 mm^3 pipette (preferably 2), micro burette, tray with tissues.

Materials. Benzenecarboxylic acid labelled with ^{14}C and unlabelled, methylbenzene, liquid scintillator (such as NE 260 supplied by Nuclear Enterprises), or dioxan containing a scintillator. (Both these liquid scintillators will take up to 10% of water without separation into two layers.)

Experimental details. Weigh out 0·12 g, approximately, of inactive benzenecarboxylic acid and dissolve in 5 cm^3 of methylbenzene. Add 10 μCi of benzenecarboxylic acid labelled with ^{14}C, or weigh out the active acid direct if of low specific activity. Pipette 1 cm^3 of the active solution into each of the four centrifuge tubes numbered 1, 2, 3 and 4, and add respectively 2, 3, 4 and 5 cm^3 of water. Cork the tubes, shake vigorously and place in the thermostat. Withdraw the tubes and shake at intervals for 15 min. Add to each scintillator phial 5 cm^3 of the special liquid scintillator, which has had nitrogen bubbled through it for 15 min in a subdued light (see Experiment 12). Mark the phials 1–9. To phials 2, 3, 4 and 5 add 50 mm^3 of the organic layer from tubes 1, 2, 3 and 4 respectively, using the micropipette, and to phials 6, 7, 8 and 9 add 100 mm^3 of the aqueous layer. There is no need to rinse the micropipettes between each addition provided the first withdrawal is made with tube No. 4. If only one micropipette is available it is advisable to start with the aqueous layer as this has the lower activity; rinse the pipette with acetone after withdrawing the four aqueous specimens. Count all nine phials, taking at least 10 000 counts for phials 2–9. Phial No. 1 should be counted for 1000 s to give the background.

The quenching effect of benzenecarboxylic acid should be examined by adding successive quantities of 100 mm^3 of the aqueous layer from tube No. 1 to phial No. 6. If necessary plot a quench correction curve. The quenching effect of methylbenzene can be examined by adding successive quantities of 50 mm^3 of methylbenzene to phial No. 2 and again plotting a quench correction curve if required. Correct all counting rates for background and quenching if necessary and divide the corrected counting rate for the aqueous layer by 2. As the corrected counting rate is proportional to the concentration of benzenecarboxylic acid, plot the log of the counting rate of the methylbenzene layer on the y axis against the log of the counting rate of the aqueous layer on the x axis and measure the slope of the graph and the intercept on the y axis.

Health hazard. There is no appreciable health hazard when using such a small quantity of ^{14}C, but all the apparatus used for the active solution should be well rinsed with a special detergent such as 'Decon'. Active work should be done over the tray. The phials of liquid scintillator should not be poured down a drain with a small trap, or blocking may be caused by precipitation of the scintillator.

Experiment 30. To investigate the effects of γ-radiolysis of an aqueous solution of benzenecarboxylic acid

Theory. γ-radiolysis of water causes the following reaction:

$$H_2O + \gamma \rightarrow H^{\cdot} + OH^{\cdot}$$

The OH radicals interact with the benzene ring forming hydroxybenzene-carboxylic acids. The products can be separated and identified by paper chromatography, scanning with a G–M tube, provided the benzene-carboxylic acid has been labelled with ^{14}C.

Apparatus. Scaler/timer with quench unit, G–M tube with thin mica end window and suitable lead castle, tall gas jar 30 cm high with cover and hook or clip attached, clynostat with a time rotation of about 15 min (not essential), recorder (if available), γ-source such as ^{60}Co of about 10 Ci strength, or X-ray apparatus. Tray with tissues.

Materials. Saturated solution of benzenecarboxylic acid, to which has been added 20 μCi of benzenecarboxylic acid labelled with ^{14}C, 0·2 cm^3, chromatography paper, butan-1-ol, 5 M ammonia solution, pyridine, saturated NaCl solution.

Experimental details. Irradiate the 0·2 cm^3 of labelled benzenecarboxylic acid solution with γ-rays, or X-rays, so that a total absorbed dose of about 2×10^{19} eV is received. The time of irradiation will depend on the strength of the γ-, or X-ray source, and the position of the specimen. The dose absorbed can be measured with the iron(II) sulphate dosimeter placed in the same position as is intended for the benzenecarboxylic acid solution. For details of this dosimeter the student is referred to standard books on radiation chemistry such as Spink and Woods, published by John Wiley, p. 105, *et seq.* The '*G*' value for the production of 1-hydroxybenzene-carboxylic acid is 0·7. (The '*G*' value is defined here as the number of molecules produced per 100 eV absorbed.)

Shake 50 cm^3 of butan-1-ol with 5 M ammonia for about ten minutes and then separate the upper (organic) layer. Cut a strip of chromatography paper 2–3 cm wide of such a length that, when attached to the hook or clip on the gas jar cover, the paper is just clear of the bottom of the jar. Make a pencil line about 2 cm from the bottom of the strip and spot 10–20 mm^3 (μl) of the irradiated solution on to the paper about the middle of the pencil line (see Experiment 24 (*a*) for details of the correct preparation of this spot). Pour the butan-1-ol, saturated with ammonia, into the gas jar to make a depth of 1 cm and insert the paper strip so that it just dips into the liquid with the spot well clear. Allow the chromatogram to run for 12 h, or overnight.

Remove the chromatogram, dry thoroughly, mark the solvent front, and mark out a scale on the side at 0·5 cm intervals from the spot. Make a slit in a piece of thin metal sheet about 0·5 cm wide by 2–3 cm long and arrange this under the window of the G–M tube. Place the paper strip under the tube as close to the slit as possible and scan the chromatogram at 0·5 cm intervals, starting from the spot, as in Experiment 24 (*a*). Counting times of 30 s or 1 min would be suitable. If a clynostat and recorder are available the strip should be attached to the clynostat and the output from the scaler/timer connected to the recorder, the G–M tube and slit being arranged as in Figure E.24. The time of rotation of the clynostat from the spot to the solvent front must be noted. In this way an automatic histogram can be recorded and the R_F values of the various peaks determined. The proportion of the various products can be calculated approximately by determining the areas under the peaks. Published figures for the ratios of the quantities of the three mono-hydroxybenzenecarboxylic acids produced are 1 : 2 : 3, 5 : 2 : 10 [48]. The student should remember that in electrophilic substitution of benzenecarboxylic acid the 2-product would be in considerable excess.

The R_F values as determined by the author for butan-1-ol/5 M ammonia are:

Benzenecarboxylic acid	0·50
1-Hydroxybenzenecarboxylic acid	0·24
2-Hydroxybenzenecarboxylic acid	0·87
3-Hydroxybenzenecarboxylic acid	0·27

The student should repeat the separation using different solvent mixtures. A mixture which gives an improved separation is:

Butan-1-ol 4 vols, pyridine 8 vols, saturated solution NaCl 5 vols, 0·880 ammonia 3 vols.

The author obtained six peaks with this mixture with R_F values of:

1-Hydroxybenzenecarboxylic acid	0·05
2-Hydroxybenzenecarboxylic acid	0·36
Benzenecarboxylic acid	0·50
1,3-Dihydroxybenzenecarboxylic acid?	0·67
3-Hydroxybenzenecarboxylic acid	0·73
?	0·80

The identification of the products was made by measuring the area under the peaks, and using the data quoted in [48].

Health hazard. There is little health hazard from the ^{14}C owing to the small quantity used. Active work should be done over a tray. The γ- or X-ray irradiation must be carried out by a qualified instructor or technician.

Experiment 31. To identify the sugars produced in photosynthesis using $^{14}CO_2$

Theory. Green plants in daylight absorb CO_2 and give off oxygen with the production of sugars and starch. It had always been assumed that the carbon in the products originated from the CO_2, but until ^{14}C became available there was no certainty. In this experiment a green plant is enclosed in an atmosphere of 50% CO_2 containing some $^{14}CO_2$, and exposed to bright light for a suitable time. The plant is removed and the sugars and starch labelled with ^{14}C extracted. The sugars are separated and identified by paper chromatography.

Apparatus. Scaler/timer or ratemeter, with quench unit, thin end-window G–M tube suitable for detecting ^{14}C, 10 or 20 cm^3 plastic syringe with hypodermic needle, 250 cm^3 round-bottomed flask fitted with Soxhlet extractor and thimble, reflux condenser, tall glass cylinder (about 40 cm high) with cork, aluminium sheet about 10 cm square with a 20 × 2 mm slit, tray with tissues.

Materials. Sodium hydrogencarbonate containing about 20 μCi of ^{14}C (the weight required is 0·04 g for a 20 cm^3 syringe and 0·02 g for a 10 cm^3 syringe), 2,3-dihydroxy-butanedicarboxylic acid (tartaric acid), 2 M sodium hydroxide solution, sprig of mint or tomato seedling, strip of chromatography paper of length to suit the cylinder, butan-1-ol, ethanoic (acetic) acid, ethanol.

Experimental details. Cut a slit in the rubber plunger of the syringe and fix the stem of the seedling into the slit. The seedling must be less than half the length of the syringe barrel. Put a few crystals of 2,3-dihydroxy-butanedicarboxylic acid into the bottom of the syringe and insert the plunger with the seedling attached. Fit the hypodermic needle to the end of the syringe and push the plunger in half way (Figure E.31).

Dissolve the ^{14}C labelled sodium hydrogencarbonate in 0·5 cm^3 or less of water and draw this solution slowly into the syringe, keeping the syringe vertical. When the reaction is complete clamp the syringe with the needle in water to act as a seal. Expose the seedling to the light of a 100 W bulb for 24 h, or expose to the sun.

Now insert the needle into some dilute sodium hydroxide solution and gently expel some of the gas; then draw in a little sodium hydroxide. Any remaining CO_2 will be absorbed. Withdraw the plunger with the seedling and test the leaves with the G–M tube held very close. There should be a considerable activity. Withdraw the seedling from the slit with a pair of forceps and place it in the thimble of the Soxhlet extractor. Attach the extractor to the round-bottomed flask which should contain about 25 cm^3 of 80% ethanol and fit the reflux condenser. Heat the flask on a boiling water bath, or with an isomantle, for an hour or two. This will extract the chlorophyll and sugars from the plant and starch will remain in the residue.

Remove the Soxhlet and evaporate the alcoholic extract to dryness on a water bath or hot plate. Place the thimble with its contents in the radioactive waste bin. Test the residue in the dish with the G–M tube. A high counting rate such as 500 counts s^{-1} should be observed. Warm the residue gently with a few cm^3 of ethanol and centrifuge to remove insoluble matter. Again evaporate to dryness.

Redissolve the residue in the minimum of ethanol—not more than 0·5 cm^3. Rule a pencil line on the chromatography paper strip about 1 cm from the end and place a minute drop of the alcoholic solution of sugars on the line, using a piece of capillary tube. Evaporate to dryness over a small flame or hair dryer. Add another drop on top of the first and evaporate again. Repeat the process two or three times. The size of the spot produced by the drop on the chromatography strip should not exceed 4 mm in diameter.

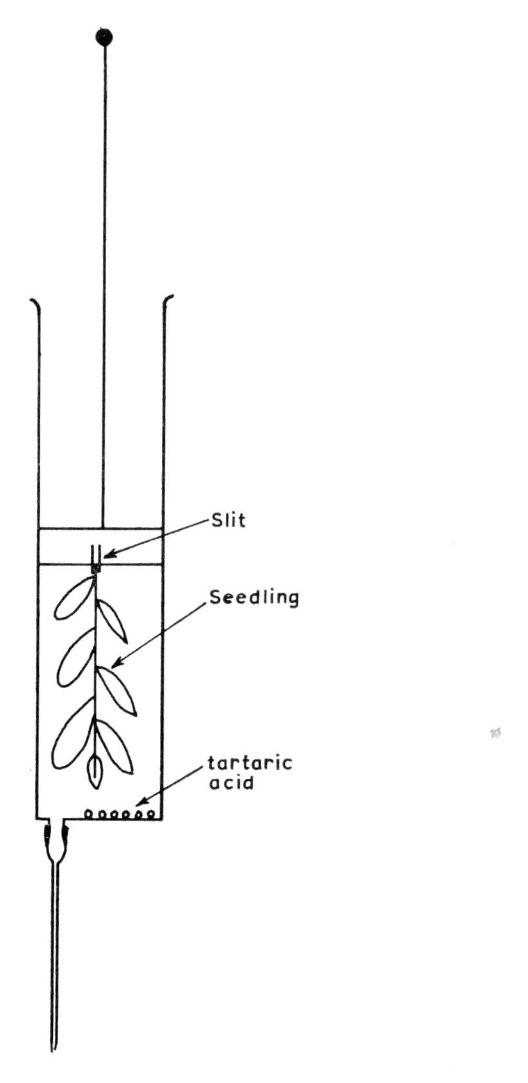

Figure E.31. This shows a suitable apparatus for producing minute quantities of labelled ^{14}C by photosynthesis. It consists of a plastic syringe, capacity 10 or 20 cm^3, with a hypodermic needle attached. A small seedling is attached to the rubber plunger of the syringe and $NaH^{14}CO_3$ solution drawn up through the needle. $^{14}CO_2$ is generated by the tartaric acid and the syringe and seedling exposed to bright light for 24 h. The sugars and starches are extracted from the seedling and separated by paper chromatography.

Make up a mixture of 40 : 10 : 50 (by volume) butan-1-ol–ethanoic acid–water and pour into the tall cylinder; about 10 cm³ should be sufficient, depending on the size of the cylinder. Suspend the paper strip in the cylinder so that its lower end just dips into the liquid, the level of which must be well below the drop on the pencil line. Stopper the cylinder and leave until the solvent front has nearly reached the top. This will take some hours depending on the grade of paper in use. A smear of silicone oil just below the top of the paper will prevent the solvent from overrunning. Mark the position of the solvent front and dry the strip thoroughly. Place the strip on a clean glass surface and find the position of the radioactive sugars by scanning with the G–M tube. The exact position of the sugars can be found by placing the aluminium sheet with the 2 mm slit in it in front of the counter and making a series of counts at varying distances from the pencil line (see Experiment 24 (*a*), p. 213). Mark the position of the sugars on the paper and measure the distances from the pencil line. Measure the total distance moved by the solvent front and calculate the R_F values (distances moved by the sugars divided by the distance moved by the solvent front) which enables the sugars to be identified. R_F values for sugars:

sucrose 0·14 fructose 0·25 glucose 0·18

Health hazard. There is no appreciable health hazard in this experiment. The active operations should be carried out over a tray, and the paper strip put in the active waste bin.

Experiment 32. To study photosynthesis with varying light exposures using *Chlorella,* detecting the products by autoradiography

Theory. This is the classical experiment of Calvin. *Chlorella,* grown in a suitable culture medium, is suspended in a buffer solution at pH5 and illuminated strongly. A suitable quantity of $NaH^{14}CO_3$ is injected into the buffer and photosynthesis allowed to proceed for a known time. The buffer, containing the *Chlorella*, is then rapidly poured into boiling 80% ethanol, which kills the *Chlorella*. The solution is evaporated to dryness, the residue is dissolved in 1 cm³ of ethanol or propanone, and spotted on to a sheet of chromatography paper. A two dimensional chromatogram is developed, dried and placed in contact with a large sheet of X-ray film. The paper is left in contact with the film for a time depending on the activity obtained. The results will depend on the time during which photosynthesis proceeded in the presence of ^{14}C. Any time longer than 3 min at room temperature will produce end products only. Exposures of the order of 10 or 30 s will produce interesting intermediate products (see Figures E.32 (*a*) and (*b*)) but quantities of ^{14}C of the order of 200 μCi will be required.

Apparatus. Autoclave, 2 Dreschel bottles, gas cylinder of air containing 1% of CO_2 (preferable), centrifuge, rotary evaporator (preferable), thin end window G–M tube and scaler or ratemeter, sheets of chromatography paper, chromatography tank, X-ray film (fast, such as Kodak Royal) at least 25×20 cm.

Materials. Culture solution and buffer at pH 5, *Chlorella* on agar slope' $NaH^{14}CO_3$ about 1 mCi, phenol, butan-1-ol, ethanoic (acetic) acid.

Experimental details. If no *Chlorella* culture is available in bulk this must be prepared by adding a commercial preparation to a sterilized culture medium, Table 32.1, in a sterilized 'Dreschel' bottle. The tubes of the bottle should be lightly plugged with cotton wool to prevent the entry of bacteria. Air, containing 1% of CO_2 from a cylinder should be gently bubbled through the culture medium, which should be exposed to a bright light. If an air cylinder with 1% of CO_2 is not available, an alternative is to suck air through a saturated solution of $NaHCO_3$, using a water pump. A suitable growth is 1% and this should be obtained in about 1 month depending on the temperature and light intensity. The cell density should be checked by centrifuging a small portion in a piece of narrow bore tubing, sealed at one end.

Portions of the *Chlorella* culture can be stored indefinitely on agar 'slopes' for future use. The agar slope is prepared from the concentrations given in Table 32.2. The mixture must be heated gently until the agar dissolves 2 or 3 cm^3 are poured into a series of sterilized test-tubes, set at an angle, to form the 'slope', and plugged with cotton wool. When cool a small amount of the *Chlorella* culture is transferred to each tube using a sterilized glass rod, or wire loop. The tubes should be stored in a refrigerator.

To carry out the photosynthesis experiment, pour about 10 cm^3 of the *Chlorella* culture, as prepared above, into a sterilized centrifuge tube; centrifuge and wash the *Chlorella* thoroughly to remove salts. Suspend the *Chlorella* in 10 cm^3 of the buffer at pH 5. The concentration of the buffer must be about 10^{-4} M. If it is less than this the starches will be difficult to remove as they will become colloidal. More concentrated buffer solutions will overload the chromatography paper with inorganic material and produce 'streaming'.

Incubate the *Chlorella* suspension by gently bubbling in air and expose to a bright light such as a photoflood bulb. After a few minutes inject a suitable quantity of $NaH^{14}CO_3$ and agitate with the air stream. For a first experiment 5 or 10 μCi would be sufficient with an exposure time of 5 min, from the moment of introducing the $NaH^{14}CO_3$. When the exposure time is complete pour the suspension rapidly into 40 cm^3 of boiling ethanol.

Evaporate the solution on a water bath. When the volume has been reduced to about 3 cm^3, add about 20 cm^3 of 20% ethanol and evaporate to approximately 5 cm^3. Centrifuge to remove the chlorophylls and starches and evaporate to dryness. Redissolve the residue in 1 cm^3 of ethanol, or acetone and spot 0·1 cm^3 in stages, on to a large sheet of chromatography paper, about 3 cm from one corner. Take great care not to allow the spot to exceed 3 mm in diameter. Do not use more than 0·1 cm^3 of the solution, or the paper will be overloaded. The preparation of the spot must not be hurried and the paper must be dried between successive additions. Count the activity of the spot by placing a thin end window G–M tube immediately above it and connecting the tube to a scaler or ratemeter. The counting rate should be about 10^2 counts s^{-1}.

Run a two-dimensional chromatogram using phenol 39 g/water 100 cm^3 for the first dimension and butan-1-ol 74 cm^3/ethanoic acid 19 cm^3/water 80 cm^3 for the second dimension. Dry thoroughly between each run and mark the solvent front. Remove the paper after the second chromatogram, again dry thoroughly and scan with the G–M tube to discover the position of the various activities. Choose a piece of X-ray film which will cover the active area of the chromatogram, including the spot, and place in contact with the paper in the dark room. A suitable green safelight must be used. Wrap the film and paper in black paper and put a weight on top. With a count of 10^2 counts s^{-1} an exposure time of 2 weeks would be sufficient for normal X-ray film, or 1 week for fast film. At the end of this time open the packet in the dark room and make a cut with a pair of scissors in the edge of film and paper so that the position of the spots on the autoradiograph can be marked on the paper. Develop the film according to the maker's instructions. Measure the R_F values of the various spots for each dimension.

To study the intermediate products of photosynthesis exposure times of as short as 5 s should be employed, but 30 s will give an interesting result. 100 or 200 μCi of ^{14}C must be used and as bright a light as possible. The evaporations must be done under vacuum without heating, to avoid hydrolysis of the intermediate products. A rotary vacuum evaporator speeds the process.

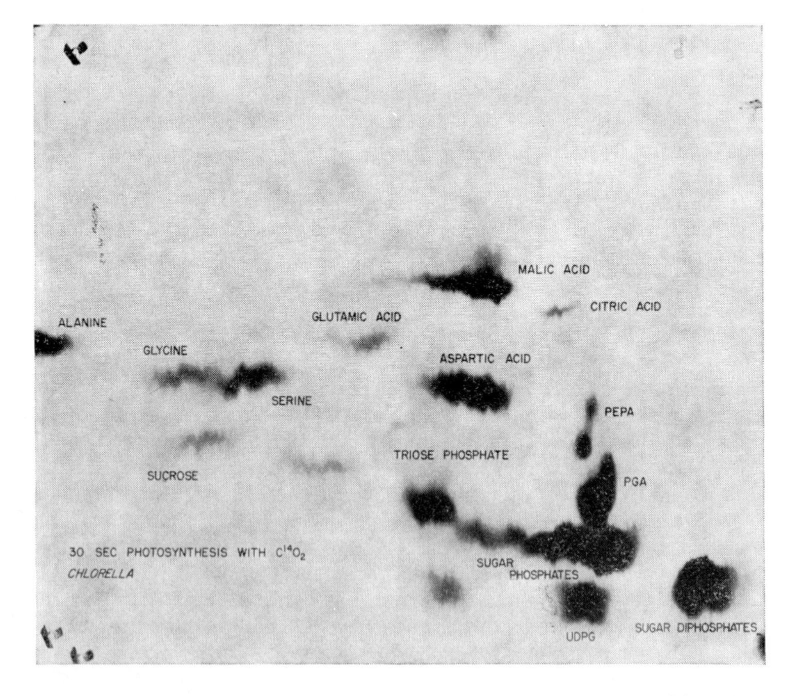

Figure E.32. (*a*) Autoradiograph showing the products of a 30 s photosynthesis with *Chlorella* and $^{14}CO_2$. (Photograph supplied by Dr. J. A. Bassham, who retains the copyright.)

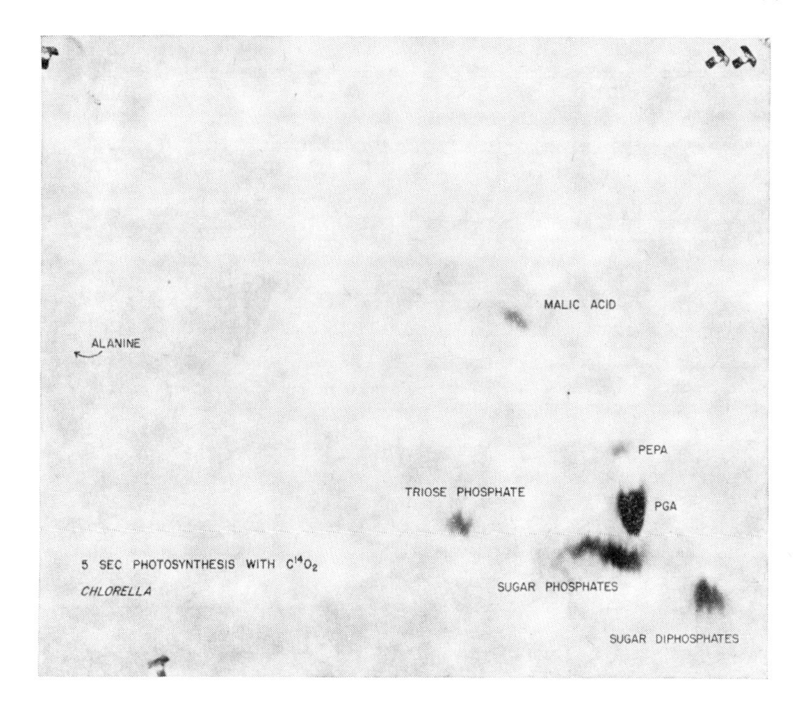

(*b*) Autoradiograph showing the products of a 5 s photosynthesis with *Chlorella* and $^{14}CO_2$. (Photograph supplied by Dr. J. A. Bassham, who retains the copyright.)

Health hazard. There is no appreciable health hazard with these quantities of ^{14}C. Active work should be done over a tray and all apparatus used for ^{14}C well rinsed with Decon. The chromatography paper should be well washed in the sink before being disposed of as solid waste.

TABLE E.32.1.

Liquid cultures	
KNO_3	1·25 g
KH_2PO_4	1·25 g
$MgSO_4.7H_2O$	2·5 g
'Trace' solution	1 cm³
Fe^{2+}/EDTA solution	1 cm³
Water	1 dm³

TABLE E.32.2.

Agar slopes	
KNO_3	0·5 g
KH_2PO_4	0·07 g
K_2HPO_4	0·09 g
$MgSO_4.7H_2O$	0·5 g
$Ca(NO_3)_2$	0·004 g
'Trace'	1·0 cm^3
Fe^{2+}EDTA	1·0 cm^3 (≡ 5 mg Fe^{2+})
Agar	15 g
Water	1 dm^3

TABLE E.32.3.

'Trace'	
$MnCl_2.4H_2O$	1·81 g
H_3BO_3	2·81 g
$ZnSO_4.7H_2O$	0·22 g
$CuSO_4.5H_2O$	0·04 g
$(NH_4)_6Mo_7O_{22}$	0·2 g
Water	1 dm^3

TABLE E.32.4.

Fe^{2+}/EDTA	
EDTA	26·1 g
$FeSO_4.7H_2O$	24·9 g
Water	1 dm^3

Experiment 33. To study the uptake of phosphate ions by plants using ^{32}P

Theory. Plants absorb salts from the soil, or culture solutions, through their roots. If a suitable salt labelled with a radioactive isotope is added to water and the roots of the plant are immersed in this solution, the uptake of the salt can be followed with the aid of a Geiger counter, and the distribution in the stem and leaves can be studied by making an autoradiograph.

Apparatus. Ratemeter, end-window G–M tube, small glass bottle (20 cm^3) in beaker or small lead cylinder for safety, syringe fitted with hypodermic needle, tray with tissues, green safelight for X-ray film.

Materials. Tomato or other small seedling, about 20 cm high, industrial X-ray film, X-ray developer (I.D.19), 10 μCi of ^{32}P as phosphate, as a solution supplied by the Radiochemical Centre, solution of sodium phosphate (about 0·01 M).

Experimental details. Fill the small bottle to within 2 cm from the top with the sodium phosphate solution, and add 1–1·5 μCi of ^{32}P from the active solution, with the aid of a syringe fitted with a hypodermic needle. This, and all subsequent work must be done over the tray.

Take the tomato plant and free its roots from earth. Place the roots in the labelled phosphate solution and secure it in the bottle with the aid of a split cork. The small bottle should be placed in a small beaker, or a small lead cylinder. After a few hours, test the stem and the leaves with the Geiger counter. Some activity will soon be observed and after 24 h there will be a very marked activity in the leaves.

Cut off a small branch with about 5 or 6 leaves. This branch should not be more than 10 cm long. Cut a thin piece of polythene sheet such as a polythene bag to fit the X-ray film, and place the tomato leaves on the sheet. Take this into the dark room, open the X-ray film carton in the green safe light, and place the polythene sheet in contact with the X-ray film with the tomato leaves on the opposite side. Replace the film in the carton with the cardboard uppermost and leave for 24 h with a suitable weight on top. Withdraw the film in the dark room and develop and fix in the usual way. The autoradiograph obtained will show a clear distribution of the radioactive phosphorus in the stem and leaves of the tomato plant (see Figure E.33).

Figure E.33. Autoradiograph of a tomato seedling, showing the distribution of phosphorus in the vascular system. A print from the original X-ray film. 1·5 μCi of ^{32}P was used, and the X-ray film was left in contact with the seedling for 24 h. (Photograph by the author.)

At the conclusion of the experiment place the tomato plant in the solid waste bin. The solution should be poured down the drain and well flushed down with water. If less time is available it would be advisable to use a larger quantity of the radioactive isotope, but it is possible to do this experiment with only 1 μCi of radioactive phosphorus.

Health hazard. There is no appreciable health hazard, but all operations should be conducted over a tray and rubber or polythene gloves should be worn while handling the radioactive phosphorus. Since ^{32}P is a pure β-emitter there is no danger from γ radiation, and the β-particles are completely stopped by the walls of the glass bottle.

Experiment 34. To study particle tracks in photographic emulsions

(a) α-ray tracks from thorium decay products

Theory. α-particles emitted by a radioactive nucleus produce a track in a photographic emulsion which can be rendered visible by development. The length of the track is a measure of the energy of the α-particle.

Apparatus and materials. 3 in by 1 in nuclear research plate (C2 or G5), 1% solution of thorium sulphate, nitrate or ethanoate (acetate).

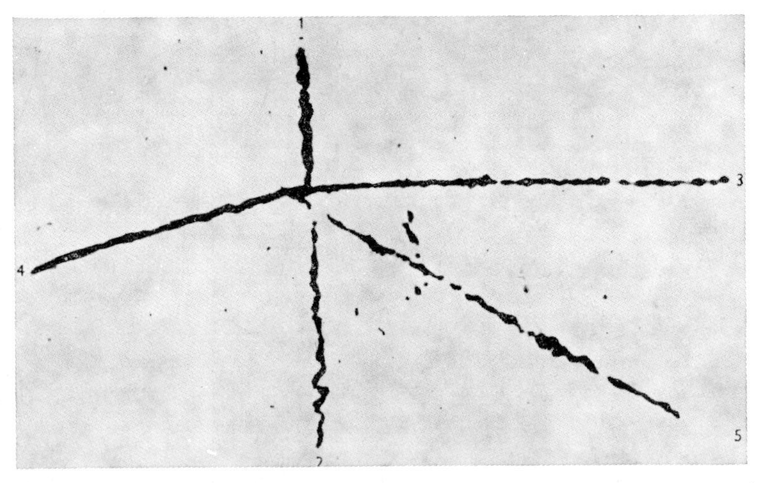

Figure E.34. (*a*) The thorium star was obtained as described in Experiment 34. Several photomicrographs were taken (five in all) using an oil immersion objective and high power eyepiece. A mosaic was constructed and this was rephotographed. As the α-particle loses energy the track becomes irregular. This is well shown by track 2. The nucleus has recoiled after the emission of particle 3. Track 3 would probably have been produced by an α-particle from ^{220}Rn and track 5 similarly from ^{212}Po. The magnification is ×3000. (Photograph made in the author's laboratory.)

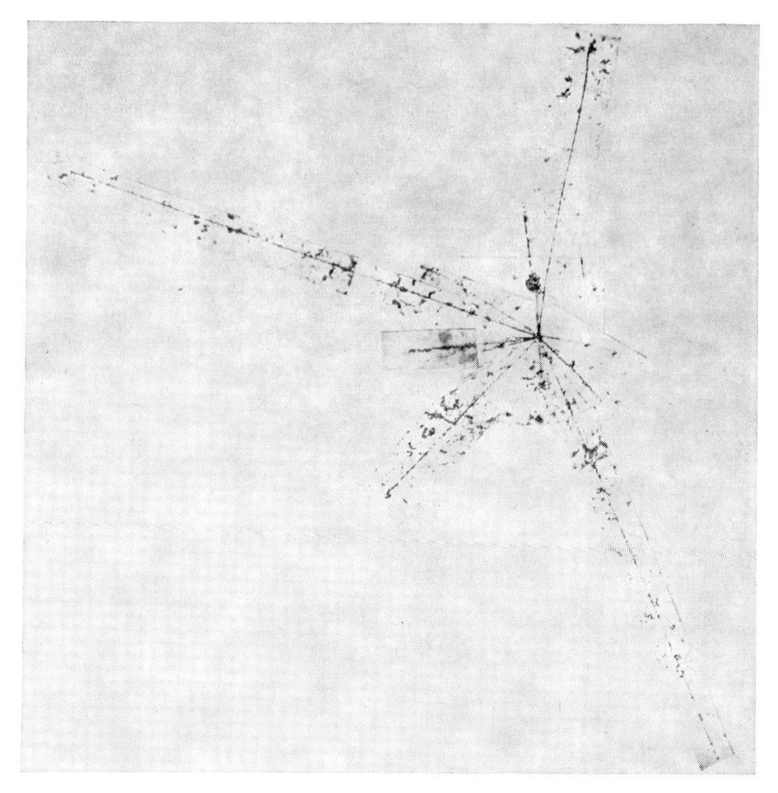

(*b*) This shows the disintegration of an Ag or Br nucleus in a photographic emulsion (Ilford G.5), which had been carried by the author in his pocket to 3000 m; it remained at this altitude for about 1 h. The tracks are produced by protons, α-particles and some heavier particles. Many of the particles escaped from the emulsion, but several did not, as can be seen from the irregular tracks which indicate that the particle is slowing down. One of the long proton tracks shows deflection. (The photomicrographs were made by two of the author's students using an oil immersion objective, high power eyepiece, and an improvised camera. Nearly 30 separate photographs had to be taken to construct the mosaic which was then rephotographed by another student. Magnification about ×2000.)

Experimental details. Soak the 3 in by 1 in nuclear research plate for 10 min in a 1% solution of thorium sulphate, nitrate or ethanoate, rinse in water and dry. Leave it in the dark for a few days and then develop as per instructions supplied with the plates. Fix thoroughly, wash well and dry. When the plate is examined under a microscope with a $\frac{1}{4}$ in or $\frac{1}{6}$ in objective and high-power eyepiece it will be found covered with stars. These stars are produced by ^{228}Th and its decay products, and the tracks are the tracks of α-particles. The stars are three-dimensional, but occasionally one will be found with all its tracks sensibly in one plane (see Figure E.34 (*a*)). A comparison of the lengths is then possible if an eyepiece with a scale is available, and from a table of range-energy relations the nuclide emitting the α-particle may be identified. Any track about double the average length will be caused by ^{212}Po. Occasionally it is possible to observe that the tracks do not all originate at the same point. This is due to the recoil of the parent nucleus. For those who are interested in photography, it is quite possible to make photomicrographs of these tracks. For the best results an oil immersion objective is required, but quite good pictures can be made with a $\frac{1}{4}$ in or $\frac{1}{6}$ in objective in conjunction with a vertical enlarger stripped of its condenser, and the light supplied with a 6 V, 30 W projector. A good film is Ilford 'line', but any fine-grain film will give satisfactory results. When the tracks are not all in one plane, a number of photographs should be taken and a mosaic constructed. This is an interesting piece of work. A range-energy table is shown below for reference:

Nuclide	Energy of α-particle in MeV	Range in emulsion in μm
^{228}Th	5·42	22
^{224}Ra	5·68	24
^{220}Rn	6·28	28
^{216}Po	6·78	32
^{212}Bi	6·09	26·7
^{212}Po	8·78	47
^{212}Po	9·45	53
^{212}Po	10·62	63·5

(*b*) *Nuclear disintegration by cosmic rays*

Theory. The primary cosmic rays coming from outer space are mainly high energy protons. They interact with atoms in the upper atmosphere and produce a complex mixture of protons, neutrons, mesons and other particles. It is possible to detect these particles at sea level, but the intensity increases rapidly with increasing altitude. When a nucleus is struck by one of these particles violent disruption may occur.

Experimental details. Take a box of 4 (3 in by 1 in) G5 or C2 Nuclear Research plates, with an emulsion at least 0·1 mm thick, to an altitude of 3000 m or higher. It is preferable, but by no means essential, to leave them at this altitude for a few weeks. The plates could equally well be sent up in a balloon, or taken on a high altitude flight.

Develop them as indicated in 34 (*a*), and scan with a $\frac{1}{4}$ in or $\frac{1}{6}$ in objective and a high power eyepiece, using a mechanical stage. One or more interesting stars should be found. Record the position on the plate by means of the mechanical stage scale. Meson tracks will also be found and with luck there may be a π, μ decay track which shows an abrupt change of direction. The meson tracks will be much fainter than those produced by protons or heavier particles.

The direction of the track of a particle may be determined by counting the grains over a given length, using an oil immersion objective. As the particle slows down the ionization produced increases (see Figure 6.1, p. 90), and so the grain density increases. Thus it is sometimes possible to determine the track of the incident particle. However, disintegrations are frequently produced by fast neutrons which produce no track.

Experiment 35. To repeat Henri Becquerel's original experiment

Theory. The β-particles given out by ^{234}Pa, which is in radioactive equilibrium with uranium, fog a photographic plate. As the β-particles are stopped by thick sheets of metal, an image of a metallic object may be obtained on the plate (see Chapter 1, p. 2).

Apparatus and materials required. X-ray film (industrial) in black carton, green safe light, X-ray developer (I.D.19), uranium compound (oxide, nitrate or chloride, 20 g).

Experimental details. Take the X-ray film in its carton and lay it in a photographic dish. Place a suitable metallic object such as a key or coin on the paper envelope marked ' tube side ', and cover with a thin sheet of polythene. Sprinkle on the polythene a thin layer of the uranium compound to make a continuous covering; the thickness need not be more than 1–2 mm. Increasing the thickness will not improve the result since the β-particles will be absorbed by thick layers. Leave for a full 24 h and develop for 8–12 min in X-ray developer. Wash and fix as usual.

Health hazard. There is no appreciable health hazard, but care should be taken to avoid spilling the uranium compound on benches, and the exposed layer of the uranium compound should be covered with a thin layer of glass, metal or perspex to prevent any damage to the eyes from the β-radiation.

Appendix 1. Average life

$$\tau = -\frac{1}{N_0} \int_{t=0}^{t=\infty} t \, dN \tag{A1.1}$$

where N_0=number of radioactive atoms of the particular species at $t=0$.

Now

$$dN = -\lambda N dt \qquad \text{(Equation 5.1)}$$

and

$$N = N_0 \exp(-\lambda t) \qquad \text{(Equation 5.4)}$$

Hence

$$dN = -\lambda N_0 \exp(-\lambda t) dt$$

Substituting for dN in Equation (A1.1) gives

$$\tau = \frac{1}{N_0} \int_{t=0}^{t=\infty} t\lambda N_0 \exp(-\lambda t) dt$$

$$= \lambda \int_{t=0}^{t=\infty} t \exp(-\lambda t) dt$$

Integrating this expression by parts:

$$= \lambda \left[\frac{t \exp(-\lambda t)}{-\lambda} \right]_0^\infty - \lambda \int_{t=0}^{t=\infty} \frac{\exp(-\lambda t)}{-\lambda} dt, \quad \text{(see Appendix 3)}$$

$$= \lambda \left[-\frac{t \exp(-\lambda t)}{\lambda} - \frac{\exp(-\lambda t)}{\lambda^2} \right]_0^\infty$$

The first term within the bracket is equal to 0 for both values, since the product of ∞ and $\exp(-\infty)$ is 0. The second term is equal to 0 and $1/\lambda^2$.

Hence

$$\tau = \lambda(0+0+0+1/\lambda^2)$$

$$= 1/\lambda$$

Appendix 2. Growth of a radioactive product

Solution of equation (5.7 (*a*)).

$$\frac{dN_B}{dt} + \lambda_B N_B - \lambda_A N_{0,A} \exp(-\lambda_A t) = 0 \tag{A2.1}$$

multiplying through by $\exp(\lambda_B t)$ gives

$$\frac{dN_B}{dt} \exp(\lambda_B t) + \lambda_B N_B \exp(\lambda_B t) = \lambda_A N_{0,A} \exp(\lambda_B - \lambda_A)t \tag{A2.2}$$

i.e.

$$\frac{d}{dt}[N_B \exp(\lambda_B t)] = \lambda_A N_{0,A} \exp(\lambda_B - \lambda_A)t$$

Integrating

$$N_B \exp(\lambda_B t) = \int \lambda_A N_{0,A} \exp(\lambda_B - \lambda_A)t \, dt$$

$$= \frac{\lambda_A N_{0,A}}{\lambda_B - \lambda_A} \exp(\lambda_B - \lambda_A)t + C \tag{A2.3}$$

when $t=0$, $N_B = N_{0,B}$

and

$$C = N_{0,B} - \frac{\lambda_A N_{0,A}}{\lambda_B - \lambda_A}$$

Hence, dividing Equation (A2.3) by $\exp(\lambda_B t)$

$$N_B = \frac{\lambda_A N_{0,A}}{\lambda_B - \lambda_A} \exp(-\lambda_A t) + N_{0,B} \exp(-\lambda_B t) - \frac{\lambda_A N_{0,A}}{\lambda_B - \lambda_A} \exp(-\lambda_B t)$$

$$= \frac{\lambda_A}{\lambda_B - \lambda_A} N_{0,A}[\exp(-\lambda_A t) - \exp(-\lambda_B t)] + N_{0,B} \exp(-\lambda_B t)$$

Appendix 3

Derivation of equations (5.18) *and* (5.19)

$$\int \exp(-x)\mathrm{d}x = -\exp(-x) \quad \text{and} \quad \int \exp(-kx)dx = -\frac{\exp(-kx)}{k}$$

Hence:

$$x = \int_{t=0}^{t=\infty} A_0 \exp(-\lambda t)\mathrm{d}t$$

$$= A_0\left[-\frac{\exp(-\lambda t)}{\lambda}\right]_{t=0}^{t=\infty}$$

$$= A_0(0 + 1/\lambda)$$

$$= A_0/\lambda$$

Similarly:

$$x_1 = \int_{t=t_1}^{t=\infty} A_0 \exp(-\lambda t)\mathrm{d}t$$

$$= A_0(0 + \exp(-\lambda t_1)/\lambda)$$

$$= A_0 \exp(-\lambda t_1)/\lambda$$

Appendix 4

Time (t_m) to maximum activity for transient equilibrium.

The activity of the mixture at time t, A_t

$$= -\left(\frac{dN_B}{dt} + \frac{dN_A}{dt}\right)$$

$$= \lambda_B N_B + \lambda_A N_A$$

From 5.8, p. 69, remembering that $N_{0,B} = 0$ at $t = 0$ and that $N_A = N_{0,A}$ exp $(-\lambda_A t)$ (Equation 5.4)

$$A_t = \frac{\lambda_A \lambda_B}{\lambda_B - c_A} N_{0,A}(\exp(-\lambda_A t) - \exp(-\lambda_B t)) + \lambda_A N_{0,A} \exp(-\lambda_A t)$$

$$\frac{d(A_t)}{dt} = \frac{\lambda_A \lambda_B}{\lambda_B - \lambda_A} N_{0,A}[-\lambda_A \exp(-\lambda_A t) + \lambda_B \exp(-\lambda_B t)] - \lambda_A^2 N_{0,A} \exp(-\lambda_A t) \qquad \text{(A4.1)}$$

$$= 0 \text{ when } t = t_m$$

Dividing by $\lambda_A N_{0,A}$ and multiplying by $(\lambda_B - \lambda_A)$ gives:

$$\lambda_B[-\lambda_A \exp(-\lambda_A t) + \lambda_B \exp(-\lambda_B t)] - (\lambda_B - \lambda_A)\lambda_A \exp(-\lambda_A t) = 0$$

(where $t = t_m$); hence

$$-\lambda_A \lambda_B \exp(-\lambda_A t) + \lambda_B^2 \exp(-\lambda_B t) - \lambda_A \lambda_B \exp(-\lambda_A t) + \lambda_A^2 \exp(-\lambda_A t) = 0$$

Collecting terms:

$$\lambda_B^2 \exp(-\lambda_B t) - \exp(-\lambda_A t)(2\lambda_A \lambda_B - \lambda_A^2) = 0$$

Rearranging:

$$\lambda_B^2 \exp(-\lambda_B t) = \exp(-\lambda_A t)(2\lambda_A \lambda_B - \lambda_A^2)$$

Therefore:

$$\frac{\lambda_B^2}{2\lambda_A \lambda_B - \lambda_A^2} = \frac{\exp(-\lambda_A t)}{\exp(-\lambda_B t)} = \exp(\lambda_B - \lambda_A)t$$

Taking ln of both sides:

$$\ln\left[\frac{\lambda_B^2}{2\lambda_A \lambda_B - \lambda_A^2}\right] = (\lambda_B - \lambda_A)t$$

Hence:

$$t = t_m = \frac{1}{\lambda_B - \lambda_A} \ln\left[\frac{\lambda_B^2}{2\lambda_A \lambda_B - \lambda_A^2}\right]$$

Example: the $^{212}Pb/^{212}Bi$ parent/daughter pair.

Maximum activity of the mixture

$$^{212}\text{Pb},\ t_{\frac{1}{2}}=10{\cdot}6\ \text{h},\quad \lambda=\frac{0{\cdot}693}{10{\cdot}6}=0{\cdot}0654\ \text{h}^{-1}$$

$$^{212}\text{Bi},\ t_{\frac{1}{2}}=1{\cdot}008\ \text{h},\ \lambda=\frac{0{\cdot}693}{1{\cdot}008}=0{\cdot}6875\ \text{h}^{-1}$$

Let A stand for ^{212}Pb, and B stand for ^{212}Bi.
Let t_m=time in h to maximum activity of the mixture after separation of the daughter ^{212}Bi.
Then:

$$t_m=\frac{1}{\lambda_B-\lambda_A}\ln\left[\frac{\lambda_B^2}{2\lambda_A\lambda_B-\lambda_A^2}\right]$$

$$=\frac{1}{0{\cdot}6221}\ln\left[\frac{0{\cdot}4727}{0{\cdot}0899-0{\cdot}0043}\right]$$

$$=\frac{1}{0{\cdot}6221}\ln\left[\frac{0{\cdot}4727}{0{\cdot}0856}\right]$$

$$=\frac{1}{0{\cdot}6221}\ln 5{\cdot}522$$

$$=\frac{1{\cdot}7091}{0{\cdot}6221}=2{\cdot}747\ \text{h}$$

Maximum activity of the daughter (^{212}Bi), independent of the activity of the parent.

Let t_m'=time to maximum activity of the ^{212}Bi.

Then the equation for time to maximum activity of the mixture simplifies to:

$$t_m'=\frac{1}{\lambda_B-\lambda_A}\ln\frac{\lambda_B}{\lambda_A}$$

(since the last term of Equation (A4.1) vanishes)

$$=\frac{1}{0{\cdot}6221}\ln\frac{0{\cdot}6875}{0{\cdot}0654}$$

$$=3{\cdot}782\ \text{h}$$

Appendix 5.
Identification of the α-particle

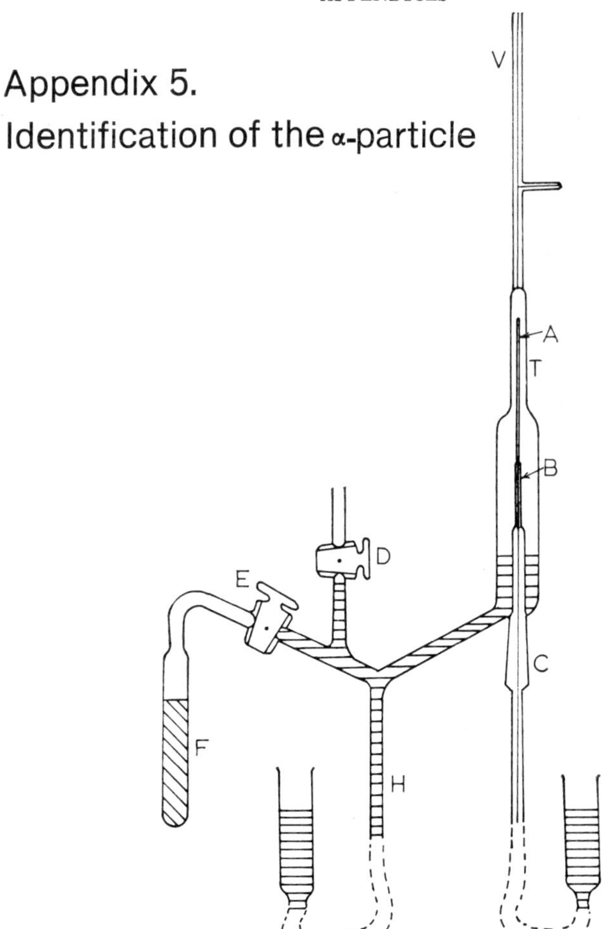

Figure A.1. Rutherford and Royds's apparatus for the identification of the α-particle. A is a fine glass tube of wall thickness less than 0·01 mm. B is a larger capillary tube which was attached to A. 140 mg of radium were enclosed in an evacuated tube for a month to reach equilibrium with radon. The radon, occupying 10^{-4} cm^3 at STP, was pumped off, purified and compressed into A. A fluorescent screen, placed close to A, glowed, showing that α-particles were passing through the glass. The tubes, A, B, were then surrounded with another glass tube T, to which was attached a small spectrum tube V. The tube T was exhausted through the taps D and E to a charcoal vacuum, by means of the charcoal in F. Any gas forming in T could be compressed into V by means of the mercury column H. After 2 days the spectrum tube showed the yellow line of helium, and after 6 days the whole helium spectrum was observed.

In order to prove that helium could not diffuse through the glass, the tube A was pumped out and pure helium was introduced. A fresh outer tube T and fresh mercury were used. After 8 days no trace of the helium spectrum was observed.

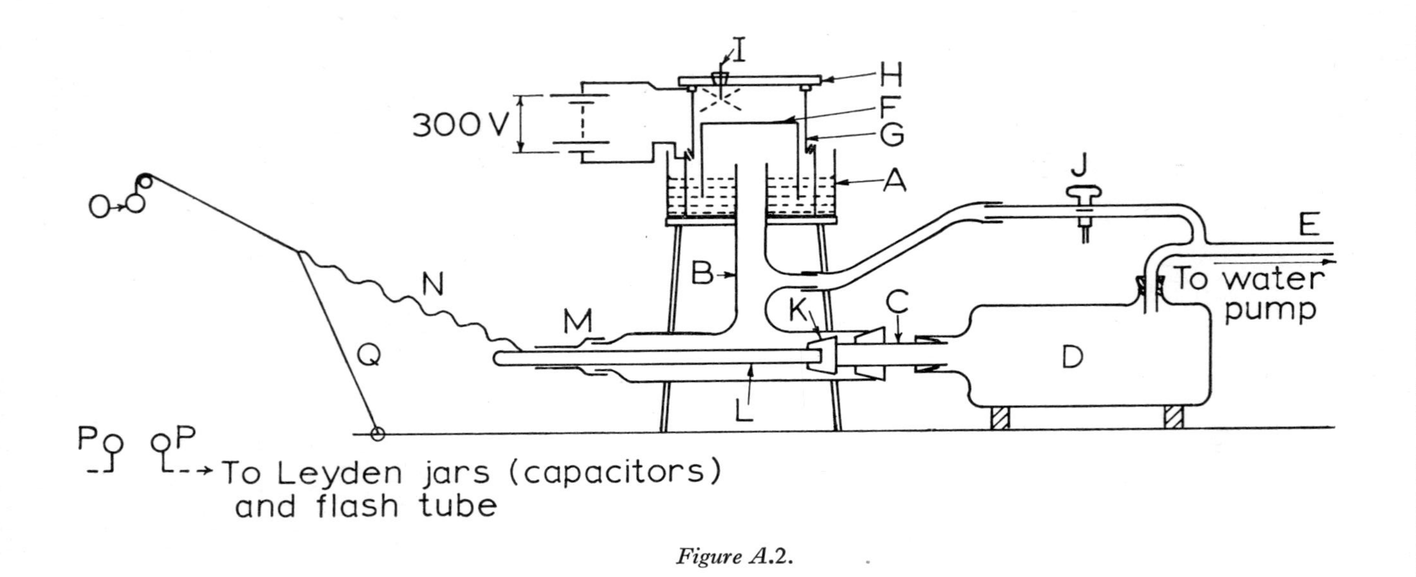
I
H
F
G
300 V
A
J
E
O
N
B
K
C
To water pump
M
Q
D
L
P
P
To Leyden jars (capacitors) and flash tube

Figure A.2.

Appendix 6. Cloud chambers

Figure A.2. The Wilson Cloud Chamber. This is the original Cloud Chamber, invented by C.T.R. Wilson in 1910. It consists of a glass or metal trough A, with a thick rubber sheet on the base, through which passes the wide bore T-piece B. The trough A is approximately half-filled with water. The T-piece may be connected, via the glass tube C, with the evacuated bottle D. A glass rod L is attached to the wide tube with a piece of rubber tubing M. A rubber cork K, attached to L, presses against the end of C, which is ground smooth to make an air-tight joint with K. F is a metal can, floating on the support of the air in B. The top of F is blackened; Wilson used Agar impregnated with indian ink. F is surrounded with a cylinder G, the base being of metal with a glass cylinder sealed on. G is closed with a disc of plate glass H, and a metal ring is fixed between G and H. A hole is drilled in H to take an α-source, I, which may be prepared from a thorium 'cow' (see Experiment 17 and Figure E.17, p. 197). A parallel beam of light (not shown) passes through the chamber and a high potential is applied as shown.

To operate, close the three-way tap J and evacuate D; a manometer should be included in the vacuum line. Connect D with the underside of F through the tap J and then readmit air. Repeat this operation several times to clear any dust particles and surplus ions from the chamber, and then pull L sharply to make a rapid expansion. Tracks will be seen coming from I, provided the expansion ratio is correct. This must be adjusted to 1·33 : 1 by altering the water level in A.

The tracks take one-fortieth of a second to form after the expansion. In order to take photographs, Wilson devised the following arrangement. A camera was placed vertically above the chamber, and a flash discharge tube at right angles. This flash tube was connected, via the brass balls PP, to a battery of Leyden jars (capacitors), charged through a Wimshurst machine turned by an assistant. Wilson attached a string N to the glass rod L, the string passing over a pulley with a lead weight attached to the end. A second string Q was attached to the first string N, so that a suitable length of slack was left in N. When the string Q was severed (Wilson generally burnt through it with his cigarette!) the weight O started to fall, tightened the string N, pulling out the glass rod sharply and making the expansion. The shock was sufficient to break the string N. The weight continued to fall, the length of the fall being adjusted so that one-fortieth of a second after the expansion it struck the brass balls PP, connecting the Leyden jars to the flash tube, which fired the flash and took the photograph.

Figure A.3. Blackett Cloud Chamber. This was produced by P. M. S. Blackett in 1932. The modification shown in the figure was designed by the author. A is a disc of plate glass 7 mm thick. B is a metal ring cemented to the disc. C is a glass cylinder. D is a metal disc with a honeycomb of holes, about 1 mm diameter, drilled through it; it is covered with black velvet. Its function is to reduce turbulence in the chamber following an expansion. E is a rubber disc, 1 mm thick, which acts as a piston. It is clamped between two metal, or perspex rings. Two aluminium plates are bolted together on either side of the disc, and the holes sealed.

F is a glass or perspex cylinder. G is a circular wooden block, with a rubber pad on top; it reduces the volume of air space below the piston, and also acts as an adjustable stop. H are three brass thumb screws (two only shown) which adjust the height of the block G and hence the expansion ratio. I is a brass plate, drilled to take 4, 15 cm bolts which, together with the brass ring L hold the chamber together. A brass cylinder J is welded to I. The apparatus is made airtight by means of five 'O' rings.

A metal disc with a rubber seating, acts as a valve. It is attached to a steel rod, screwed to a soft iron bar K, and passes through a well fitting bush in J. The valve is released by a coiled spring and held closed by the electromagnet M.

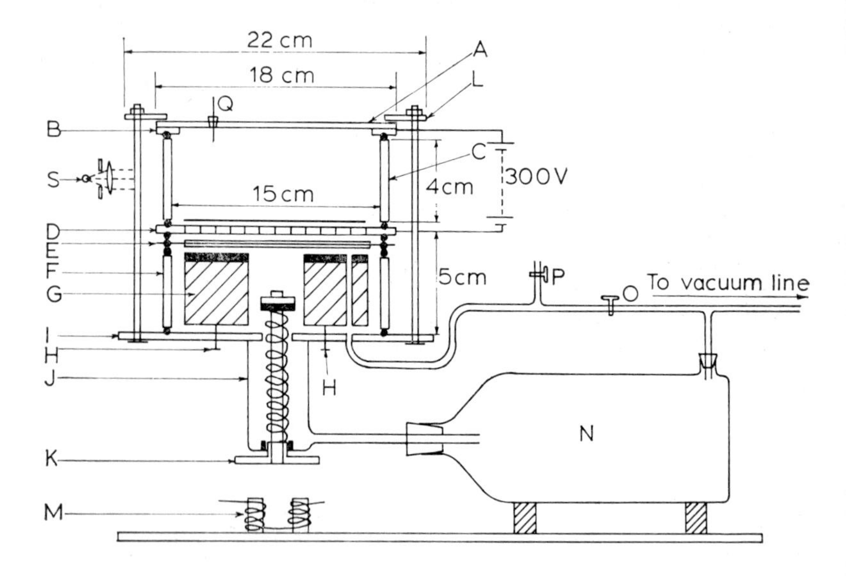

Figure A.3.

To operate the Cloud Chamber, put a few drops of 50% methanol/water into the upper chamber. Insert the α-source Q (see Wilson Cloud Chamber). Press K down into contact with the electromagnet. Close the tap P, and open O slowly to draw down the piston E. Open P to readmit air. Repeat the operation several times to remove dust particles and excess ions. Switch off the electromagnet to make a rapid expansion. The thumb screws H must be adjusted until tracks become visible.

To take photographs, the light source must be replaced with a flash tube. An electronic delay circuit, to delay the flash 1/40 s after the expansion, must be connected through the switch which controls the electromagnet, to the flash tube. The photograph, Figure A.4, was taken with the author's apparatus.

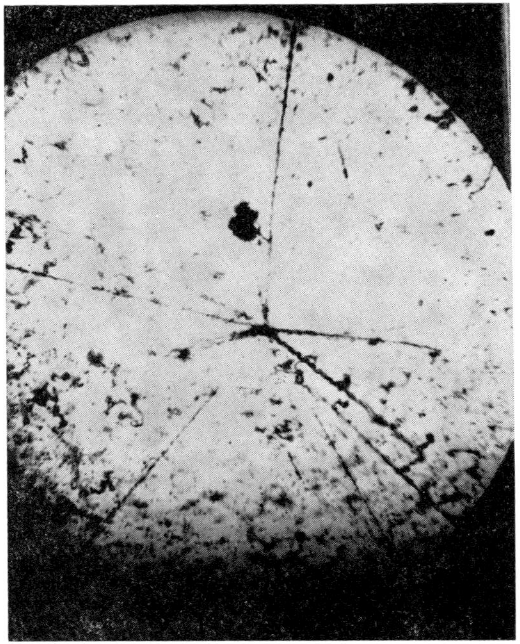

Figure A.4. *α-particle tracks*. Photograph taken by Rev. J. L. Birley with the author's apparatus (Figure A.3), using a flash tube with a 1/40ths delay after the expansion. The source of the α-particle tracks is ^{212}Pb and its decay products, on a copper wire charged from a thorium ' cow '. The long tracks are produced by ^{212}Bi and ^{212}Po.

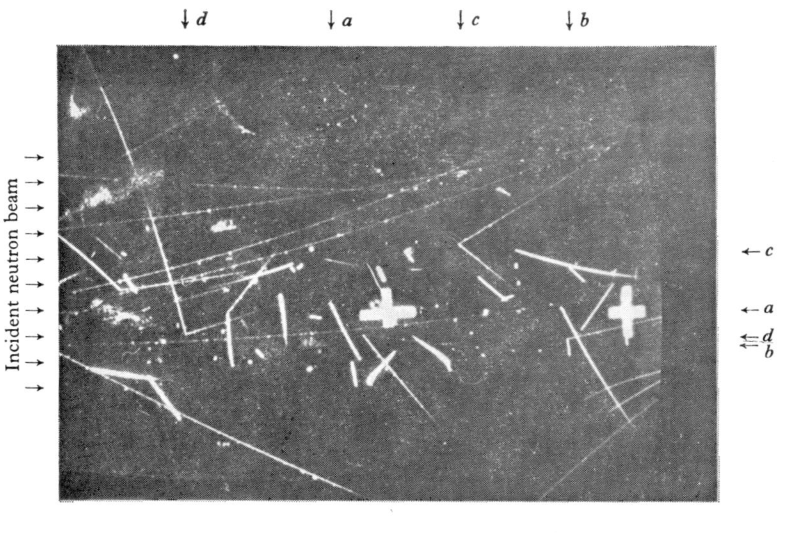

Figure A.5. A Diffusion Cloud Chamber. Photograph showing the disintegration of nuclei of helium atoms by high energy neutrons. The following reactions may be seen:

a $n + {}^{4}He \rightarrow n + n + {}^{3}He$

b $n + {}^{4}He \rightarrow n + {}^{1}H + {}^{3}H$

c $n + {}^{4}He \rightarrow {}^{2}H + {}^{3}H$

d $n + {}^{4}He \rightarrow n + n + {}^{1}H + {}^{2}H$

The incident and recoil neutrons are invisible since they carry no electrical charge and hence produce no cloud chamber track. The long fine tracks crossing the chamber are due to protons ejected from the side wall of the cloud chamber by neutrons in the incident beam.

[Photograph by Dr. D. F. Shaw].

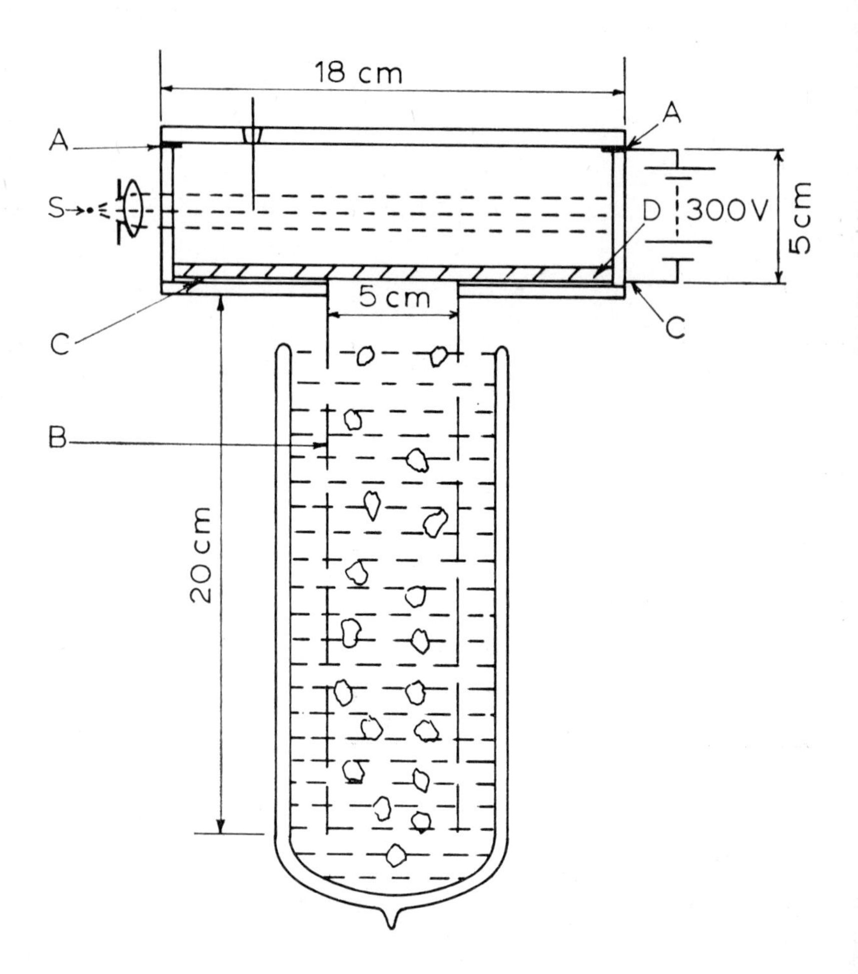

Figure A.6. Continuous Cloud Chamber. This is by far the simplest to construct, but it requires a supply of 'dry ice' or liquid nitrogen. It consists of a glass or perspex box (the author made one of perspex), of the dimensions shown. A metal ring, A, is sealed under the lid. A metal tube B passes through the base (sealed) and is connected to a metal disc C. On top of the metal disc is a sheet of black felt D. The metal ring and disc have connections for the high voltage supply.

To operate the chamber, add about 1 cm^3 of 50% methanol/water through the hole in the lid, insert the α-source (see Wilson Cloud Chamber) and leave for 20–30 min for the chamber to become saturated with the vapour. Place the metal tube B in a mixture of dry ice and propanone, or liquid nitrogen and methylbenzene, in a darkened room, connect the high voltage supply as indicated, and switch on the parallel beam of light from the source S. Tracks should appear within a few minutes.

Appendix 7. The construction of castles for G-M tubes

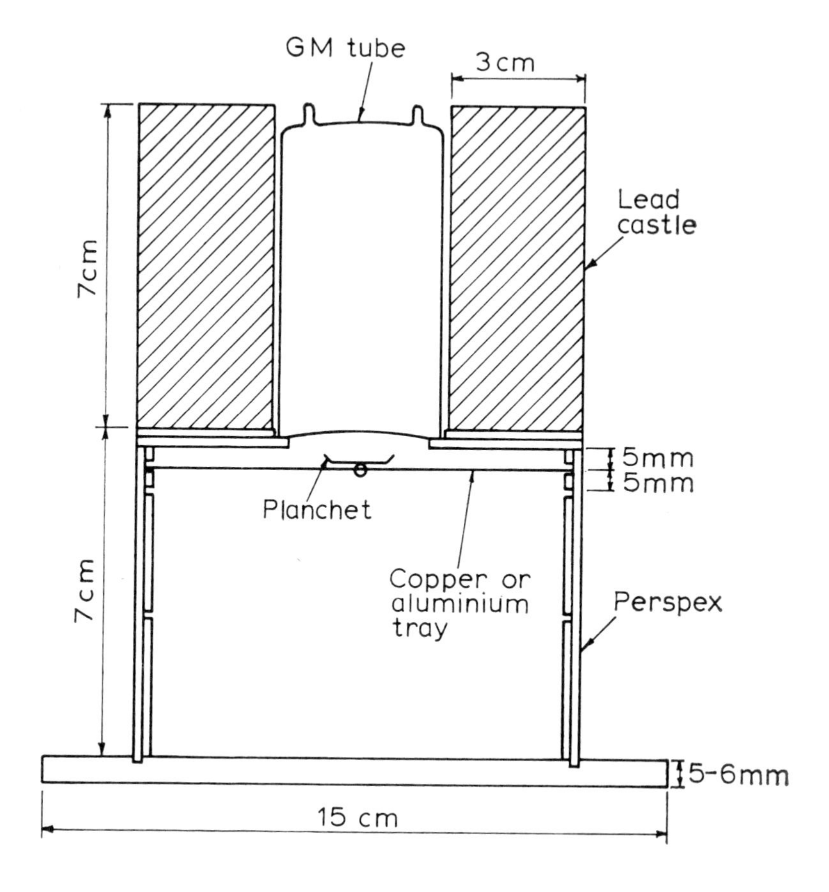

Figure A.7. End window tube. The tube holder is conveniently made from perspex using chloroform (trichloromethane) as fixative. The dimensions are marked on the figure. The sides and top should be 6 mm thick, or may be made in two pieces of 3 mm. The back should be about 6 mm thick and detachable, so that strips of chromatography paper may be slid through the holder. The lead castle may be cast from scrap lead in a tin, using a metal tube, of suitable diameter to suit the G–M tube, as a former.

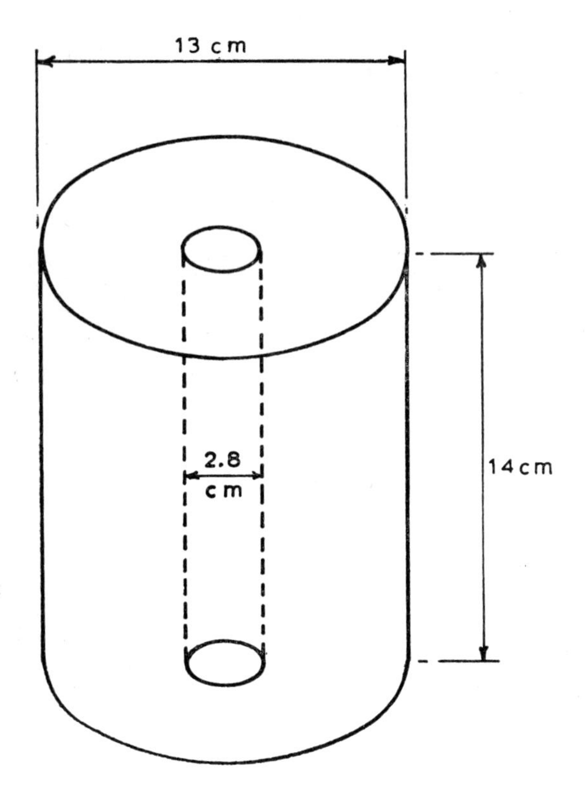

Figure A.8. Lead castle for liquid counting G–M tube. The castle may be made as for the end window tube. It may be supported on a strong tripod, and the G–M tube held in position by means of a plastic terry clip. This castle is suitable for a G–M tube with a two-pin base.

Appendix 8. Suppliers of radioisotopes

The following are the addresses of the principal suppliers:

(1) Great Britain	The Radiochemical Centre, White Lion Rd., Amersham, Bucks HP7 9LL.
(2) USA	Amersham Corporation, 2636 South Clearbrook Drive, Arlington Heights, Illinois 60005. New England Nuclear Corporation, 549 Albany St., Boston, Massachussetts 02118.
(3) Germany	Amersham Buchler GmbH and Co KG, Harxbutteler Strasse 3, D–3301 Wenden Uber, Braunschweig.
(4) India	The Bhabha Atomic Research Centre, Trombay, Bombay 400085.
(5) Australia	The Australian Atomic Energy Commission, New Illawarra Road., Lucas Heights, NSW 2232.

Appendix 9. Physical constants and units

curie (Ci)	$3{\cdot}7 \times 10^{10}\ s^{-1} = 3{\cdot}7 \times 10^{10}$ Bq
becquerel (Bq)	s^{-1}
rad	$10^{-2}\ J\ kg^{-1} = 10^{-2}$ Gy
gray (Gy)	$J\ kg^{-1}$
barn (b)	$10^{-28}\ m^2$, or $10^{-24}\ cm^2$
Unified atomic mass	$1{\cdot}6606 \times 10^{-27}$ kg
Electron rest mass	$9{\cdot}1096 \times 10^{-31}$ kg
Proton rest mass	$1{\cdot}6726 \times 10^{-27}$ kg
Neutron rest mass	$1{\cdot}6749 \times 10^{-27}$ kg
Speed of light in vacuo	$2{\cdot}9979 \times 10^{8}\ m\ s^{-1}$
Avogadro constant	$6{\cdot}0220 \times 10^{23}\ mol^{-1}$
Energy equivalent of 1 a.m.u.	$1{\cdot}4924 \times 10^{-10}$ J
Energy equivalent of 1 electron mass	$8{\cdot}1871 \times 10^{-14}$ J
Energy equivalent of 1 proton mass	$1{\cdot}5033 \times 10^{-10}$ J
Energy equivalent of 1 neutron mass	$1{\cdot}5054 \times 10^{-10}$ J
Electron volt (eV)	$1{\cdot}6022 \times 10^{-19}$ J

Temperature equivalent to 1 eV $\approx 10^4$ K.

References

[1] Becquerel, H., *Compt. Rend.*, **122,** 420, 501 (1896).
[2] Curie, M., *Compt. Rend.*, **27,** 175, 1215 (1898).
[3] Debierne, A., *Compt. Rend.*, **129,** 593 (1899).
[4] Crookes, W., *Proc. Roy. Soc.* A, **66,** 409 (1900).
[5] Rutherford, E., and Royds, T., *Phil. Mag.*, **17,** 281 (1909).
[6] Wilson, C. T. R., *Proc. Roy. Soc.* A, **85,** 285 (1911).
[7] Rutherford, E., *Phil. Mag.*, **21,** 669 (1911).
[8] Bohr, N., *Phil. Mag.*, **26,** 1 (1913).
[9] Rutherford, E., *Phil. Mag.*, **37,** 581 (1919).
[10] Cockcroft, J. D., and Walton, E. T. S., *Proc. Roy. Soc.* A, **136,** 619 (1932).
[11] Chadwick, J., *Proc. Roy. Soc.* A, **136,** 692 (1932).
[12] Fermi, E., *Nature*, **133,** 898 (1934).
[13] Quill, L. L., *Chem. Rev.*, **23,** 87 (1938).
[14] Curie, J., and Savitch, P., *Compt. Rend.*, **206,** 906 (1938).
[15] Hahn, O., and Strassmann, F., *Naturwiss.*, **27,** 11, 89 (1938).
[16] Meitner, L., and Frisch, O. R., *Nature*, **143,** 239 (1939).
[17] McMillan, E., and Abelson, P., *Phys. Rev.*, **57,** 1185 (1940).
[18] Kennedy, J. W., Seaborg, G. T., Segrè, E., and Wahl, A. C., *Phys. Rev.*, **70,** 555 (1946).
[19] Urey, H. C., *Proc. Acad. Nat. Sci.*, **18,** 496 (1932).
[20] Thompson, S. G., and Seaborg, G. T., *Progress in Nuclear Energy*, Series **3,** 3. Pergamon Press Ltd. (London) (1956).
[21] Seaborg, G. T., McMillan, E. M., Kennedy, J. W., and Wahl, A. C., *Phys. Rev.*, **69,** 366 (1946).
[22] Ghiorso, A., James, R. A., Morgan, L. O., and Seaborg, G. T., *Phys. Rev.*, **78,** 472 (1950).
[23] Seaborg, G. T., James, R. A., and Ghiorso, A., Paper 22.2, 1554 of *The Transuranium Elements*, Natl. Nuclear Energy Ser. Div. **14B** (1949).
[24] Thompson, S. G., Ghiorso, A., and Seaborg, G. T., *Phys. Rev.* **77,** 838 (1950).
[25] Thompson, S. G., Street, K., Jr., Ghiorso, A., and Seaborg, G. T., *Phys. Rev.*, **80,** 790 (1950).
[26] Ghiorso, A., *et al.*, *Phys. Rev.*, **99,** 1048 (1955).
[27] Ghiorso, A., *et al.*, *Phys. Rev.*, **98,** 1518 (1955).
[28] Seaborg, G. T., *Ann. Rev. Nuclear Sci.*, **18,** 53 (1968).
[29] Flerov, G. N., Druin, V. A., and Pleve, A. A., *Soviet Phys. Uspekhi*, **13,** 24 (1970).
[30] Silva, R., *et al.*, *Inorg. Nuclear Chem. letters*, **6,** 871 (1970).

[31] ZVARA, I., *et al.*, *J. Inorg. Nuclear Chem.*, **32,** 1885 (1970).
[32] FLEROV, G. N., *Science*, **170,** 15 (1970).
[33] GHIORSO, A., *Science*, **171,** 127 (1970).
[34] GHIORSO, A., *et al.*, *Phys. Rev. letters*, **24,** 1498 (1970).
[35] NILLSON, S. G., THOMPSON, S. G., and TSANG, C. F., *Phys. letters*, **28B,** 458 (1969).
[36] MARINOV, A., *et al.*, *Nature*, **229,** 464 (1971).
[37] MORRISON, G. H., *Analyt. Chem.*, **43,** 22A (1971).
[38] BRAUN, T., and TOLGYESSY, J., *Radiometric Titrations*, Internat. Ser. Monographs Analyt. Chem., Pergamon, Oxford, vol. 29 (1967).
[39] HEVESEY, G., *Trans. Faraday Soc.*, **34,** 841 (1938).
[40] SZILARD, L., and CHALMERS, T. A., *Nature*, **134,** 462, 494 (1934).
[41] LIBBY, W. F., *J. Am. Chem. Soc.*, **69,** 2523 (1947).
[42] PEACOCKE, T. A. H., and WALTON, G. N., Reaction of iodine in the neutron irradiation of iodobenzene, *J. Chem. Soc.*, **A,** 1264 (1969).
[43] *Nuclear Data Tables*, Consistent set of *Q* values, U.S.A.E.C. (1961).
[44] BALZER, R., *et al.*, *Helv. Phys. Acta*, **32,** 264 (1959).
[45] ESTULIN, I. V., KALINKIN, L. F., and MELIORANSKII, A. S., *Soviet Phys.*, **5,** 801 (1957).
[46] LIBBY, W. F., ANDERSON, E. C., and ARNOLD, J. R., *Science*, **109,** 227 (1949).
[47] STRUTT, L., *Proc. Roy. Soc.* A, **76,** 95 (1905).
[48] LOEBL, H., STEIN, G., and WEISS, J., *J. Chem. Soc.*, 405 (1951).

Suggestions for further reading

CATCH, J. R., *Carbon-14 Compounds*, Butterworth, London (1961).
COOK, G. B., and DUNCAN, J. F., *Modern Radiochemical Practice*, Oxford University Press (1952).
FAIRES, R. A. and PARKES, G., *Radioisotope Laboratory Techniques*, 3rd Edition, Butterworth, London (1973).
FRIEDLANDER, G., KENNEDY, J. W., and MILLER, J. M., *Nuclear and Radiochemistry*, John Wiley & Sons, Inc., New York (1955).
HEVESEY, G., and PANETH, F. A., *A Manual of Radioactivity*, Oxford (1938).
HEVESEY, G., *Radioactive Indicators*, Interscience Publishers, New York and London (1948).
HULME, H. R., *Nuclear Fusion*, Wykeham Publications (London) (1969).
LEWIS, G. M., *Neutrinos*, Wykeham Publications (London) (1970).
MCKAY, H. A. C., *Principles of Radiochemistry*, Butterworth, London (1971).
OVERMAN, R. T., and CLARK, H. M., *Radioisotope Techniques*, McGraw-Hill Book Co., New York (1960).
PEACOCKE, T. A. H., *Atomic and Nuclear Chemistry*, Pergamon (1967).
ROTHERHAM, L., Nuclear power systems, *Contemp. Phys.*, Vol. 15, No. 5 (1974).
WHITEHOUSE, W. J., and PUTMAN, J. L., *Radioactive Isotopes*, Oxford University Press (1953).

International Council for Radiological Protection, Publication 2.
A.M. 65, Department of Education and Science.
Code of Practice, H.M.S.O.

Answers to numerical questions

1.1. $6{\cdot}04\times10^{23}$

1.3. (*a*) $3{\cdot}12\times10^{4}$ dis s^{-1} (*b*) $8{\cdot}76\times10^{4}$ dis s^{-1}

2.1. $4{\cdot}49\times10^{9}$ J

2.2. $8{\cdot}20\times10^{13}$ J, $2{\cdot}05\times10^{7}$ J

2.3. $2{\cdot}96\times10^{-7}$ s

4.1. 4 min

5.1. 1·07 min

5.2. 1·0098

5.3. ^{232}Th 10^{12} y, ^{216}Po $1{\cdot}5\times10^{-2}$ s, ^{212}Po $1{\cdot}2\times10^{-11}$ s

5.5. $1{\cdot}10\times10^{14}$ y

5·6. 33·04 d

5.7 (i) 31·9 μCi g^{-1} (ii) 275 μCi g^{-1}

5.8. $2{\cdot}7\times10^{5}$ y

5.9. 2·6 min

5.10. (*b*) 164 min (*d*) 15·5 min

5.11. $3{\cdot}00\times10^{3}$ kg, $3{\cdot}27\times10^{5}$ Ci

10.1. Minimum quantities of ^{131}I and ^{110m}Ag required when using G–M tube, 350 and 700 μCi respectively. Minimum quantities when using scintillation counter 35 μCi in both cases.

10.2. S +5, T 0, U −1, W +1

10.3. 1·14%

11.1. 187 eV

12.1. (*a*) $2{\cdot}69\times10^{-4}$ rem y^{-1} (*b*) $1{\cdot}49\times10^{6}$ rem y^{-1}, $3{\cdot}3\times10^{4}$ rem y^{-1}

12.2. (i) (*a*) $2{\cdot}7\times10^{-1}$ (*b*) $3{\cdot}4\times10^{2}$

(ii) (*a*) γ $3{\cdot}78\times10^{-2}$, β $3{\cdot}5\times10^{-2}$ (*b*) γ 47, β 59

(iii) (*a*) 6×10^{-3} (*b*) 31

Time (i) (*a*) 3·7 min (*b*) 0·18 s

(ii) (*a*) 5·6 min (*b*) 0·25 s

(iii) (*a*) 2·5 h (*b*) 2 s

Index of names

The numbers in parenthesis refer to references. Numbers in bold type are principal discussions of research.

Subject index

(Page numbers beyond 167 refer to experimental section)